Birkhäuser

João Pedro Morais
Svetlin Georgiev
Wolfgang Sprößig

Real Quaternionic Calculus Handbook

 Birkhäuser

João Pedro Morais
CIDMA
University of Aveiro
Aveiro
Portugal

Wolfgang Sprößig
Institut für Angewandte Analysis
TU Bergakademie Freiberg
Freiberg
Germany

Svetlin Georgiev
Department of Differential Equations
University of Sofia St Kliment
Ohridski Faculty of Mathematics and
Informatics
Sofia, Bulgaria

ISBN 978-3-0348-0621-3   ISBN 978-3-0348-0622-0 (eBook)
DOI 10.1007/978-3-0348-0622-0
Springer Basel Heidelberg New York Dordrecht London

Library of Congress Control Number: 2013957805

Mathematics Subject Classification 2010: 17C60, 30G35, 05A05, 15A66, 15A54, 51M05

Preface

Real Quaternionic Analysis is a multifaceted subject. Created to describe some phenomena in special relativity, electrodynamics, spin, etc. It has developed into a body of material that interacts with many branches of mathematics, such as complex analysis, harmonic analysis, differential geometry, differential equations, as well as into a ubiquitous factor in the description and elucidation of problems in mathematical physics. Quaternions have been towards Maxwell's equations. In the meantime, real quaternionic analysis has became a well-established branch of mathematics and greatly successful in many different directions. Quaternions have been successfully applied to signal processing, most notably pattern recognition. They can be used for image segmentation, finding structure based not only upon color, but repeating patterns. Quaternions may also be used to simplify derivations in computer vision and robotics, to develop computer applications in virtual reality, and so on.

This book is intended to provide material for an introductory one- or two-semester undergraduate course on some of the major aspects of real quaternionic analysis, with exercises. Alternatively, it may be used in a beginning graduate level course and as a reference. That it is why, rather than general theorems, we supply concrete examples and exercises which form the basis of this book. The exercises proposed at the end of each chapter are an essential part of it. The writing herein is straightforward and it is addressed to readers who have no prior knowledge of this subject and who have a basic graduate mathematics background: real and complex analysis, ordinary differential equations, partial differential equations and theory of distributions.

The detailed reference list proposed at the end is seen as an initial point for the development of the topics covered in the handbook. From a reader's point of view, we chose to present it at the end rather than throughout the text.

Here is a brief description of the topics covered in the first ten chapters.

Chapter 1   An introduction to and historical background on the discovery of the quaternions are provided. The definitions and general properties of quaternions are examined in detail.

Chapter 2   The notions of quaternion and spatial rotation are mastered. Some applications of quaternions to plane geometry are mentioned.

Chapter 3   Studies sequences of quaternion numbers and their properties.

Chapter 4   Reviews the basic properties of quaternion power series and infinite products.

Chapter 5   The quaternion exponential, logarithmic and power functions are covered. A brief discussion on the notions of multiple-valued functions and branches is also presented.

Chapter 6   The quaternion trigonometric functions are defined.

Chapter 7   The quaternion hyperbolic functions are introduced.

Chapter 8   The main focus here is on the study of the inverses of the quaternion trigonometric and hyperbolic functions, and their properties.

Chapter 9   Matrices with quaternion entries are presented. In spite of the difficulties caused by the noncommutativity of the multiplication of quaternions, we still manage to introduce the concepts of determinant, rank, eigenvalues, and relations of similarity.

Chapter 10   Studies the concepts of monomials, polynomials and binomials involving quaternion numbers.

It is with great pleasure that we express our appreciation to all those who have expressed support, enthusiasm and encouragement in this adventure. We are forever indebted to our families and close friends for their patience, understanding and support: Lucília and Mário Morais, José António Morais, Ana Beatriz Pistola, Constança Sofia Morais, Francisco Elías, Laura, Borislav, Gerard and Nathalie, and Martina Sprößig. As regards to the present edition, our thanks go to Helmuth Malonek (Aveiro/Portugal), Isabel Cação (Aveiro/Portugal), Klaus Gürlebeck (Weimar/Germany), Tao Qian (Macau/China), Mahmoud Abul-Ez (Sohag/Egypt), Kou Kit Ian (Macau/China), Eckhard Hitzer (Fukui/Japan), Saburou Saitoh (Aveiro/Portugal), Hoai Le (Freiberg/Germany), and Inês Matos (Aveiro/Portugal) for helpful discussions and encouragement, and we especially thank the editor Thomas Hempfling (Birkhäuser) for the meticulous care with which he examined the entire manuscript. We are also grateful to the students who participated in the course "Function theory in higher dimensions" at the Technical University of Mining, Freiberg (Germany), whose enthusiasm, interest and dedication are admirable. For financial aid, the first named author wishes to express his gratitude to the Portuguese Foundation for Science and Technology ("FCT–Fundação para a Ciência e a Tecnologia") via the postdoctoral grant SFRH/BPD/66342/2009. The first author's work is also supported by *FEDER* funds through *COMPETE*–Operational Programme Factors of Competitiveness ("Programa Operacional Factores de Competitividade") and by Portuguese funds through the *Center for Research and Development in Mathematics and Applications* (University of Aveiro) and the FCT, within project PEst-C/MAT/UI4106/2011 with COMPETE number FCOMP-01-0124-FEDER-022690.

Although the examples and exercises have been tested several times, we apologize in advance for any errors (typos) or just plain mistakes that you may find, and kindly ask you to bring them to our attention.

<table>
<tr><td>Aveiro, Portugal</td><td>João Pedro Morais</td></tr>
<tr><td>Sofia, Bulgaria</td><td>Svetlin Georgiev</td></tr>
<tr><td>Freiberg, Germany</td><td>Wolfgang Sprößig</td></tr>
<tr><td>November 2012</td><td></td></tr>
</table>

Contents

An Introduction to Quaternions

1

Once we start studying quaternionic analysis we take part in a wonderful experience, full of insights. This ideology is shown, for instance, when we start describing the first results and pursuing the subject, while the amazement lingers on through the elegance and smoothness of the results.

Quaternions were discovered on 16th of October 1843 by the Irish mathematician Sir William Rowan Hamilton (1805–1865). His original motivation was to create a type of hypercomplex numbers related to the three-dimensional space in the same way as the standard complex numbers are related to the plane. In his first conjecture, complex numbers would need one extra imaginary part, i.e. one real part and two distinct imaginary parts. For the generalization of the so-called "Theory of Couplets" to the "Theory of Triplets", Hamilton had fundamentally in mind the use of these triplets to represent rotations in the three-dimensional space, just like complex numbers may be used to represent rotations in the two-dimensional plane. In his later papers those triplets are termed number vectors. Therefore, it is only natural that in the attempt to achieve such a construction certain questions have remained untouched.

It should be noted that even before W.R. Hamilton, a multiplication of 4-vectors similar to quaternion multiplication, was already known to leading mathematicians such as Leonhard Euler (1707–1783), Carl Friedrich Gauss (1777–1855) and Olinde Rodrigues (1795–1851). L. Euler discovered in 1748 the four-square identity and used "quaternion" parameter representations in order to describe motions in the Euclidean space. This fact was rediscovered by Wilhelm Blaschke (1885–1962) in his speech on the occasion of the 250th birthday of L. Euler in Berlin in 1957. C.F. Gauss thought that it made no sense to tell all recent results to his contemporaries, because they would not understand the implications. In consequence, his results on quaternions remained unpublished during his life, and were only made public in 1900. O. Rodrigues described a parametrization of general rotations through four parameters. He used a special multiplication technique which constitutes an anticipation of the quaternion multiplication.

After nearly 10 years of unsuccessfully searching for a three-dimensional extension of complex numbers, W.R. Hamilton found himself on the brink of

J.P. Morais et al., *Real Quaternionic Calculus Handbook*,
DOI 10.1007/978-3-0348-0622-0_1, © Springer Basel 2014

1

giving up. He spent years trying to solve this problem without any success; even his children were aware of his frustrating attempts at these numbers. This sequence of events is documented in one of his letters to his eldest son Archibald Hamilton, which we now briefly quote:

> Letter from Sir W.R. Hamilton to Rev. Archibald H. Hamilton.
> Letter dated August 5, 1865.
> MY DEAR ARCHIBALD - (1) I had been wishing for an occasion of corresponding a little with you on QUATERNIONS: and such now presents itself, by your mentioning in your note of yesterday, received this morning, that you "have been reflecting on several points connected with them" (the quaternions), "particularly on the Multiplication of Vectors." (2) No more important, or indeed fundamental question, in the whole Theory of Quaternions, can be proposed than that which thus inquires What is such MULTIPLICATION? What are its Rules, its Objects, its Results? What Analogies exist between it and other Operations, which have received the same general Name? And finally, what is (if any) its Utility? (3) If I may be allowed to speak of myself in connexion with the subject, I might do so in a way which would bring you in, by referring to an ante-quaternionic time, when you were a mere child, but had caught from me the conception of a Vector, as represented by a Triplet: and indeed I happen to be able to put the finger of memory upon the year and month - October, 1843 - when having recently returned from visits to Cork and Parsonstown, connected with a meeting of the British Association, the desire to discover the laws of the multiplication referred to regained with me a certain strength and earnestness, which had for years been dormant, but was then on the point of being gratified, and was occasionally talked of with you. Every morning in the early part of the above-cited month, on my coming down to breakfast, your (then) little brother William Edwin, and yourself, used to ask me, "Well, Papa, can you multiply triplets"? Whereto I was always obliged to reply, with a sad shake of the head: "No, I can only add and subtract them."

But on the 16th day of the same month—which happened to be a Monday, and a Council day at the Royal Irish Academy—while walking along the Royal Canal in Dublin with his wife, W.R. Hamilton was struck by inspiration. Instead of the two imaginary parts as he had first considered, he imagined a three part imaginary system. In Hamilton's own words:

> Tomorrow will be the fifteenth birthday of the Quaternions. They started into life, or light, full grown, on the 16th of October, 1843, as I was walking with Lady Hamilton to Dublin, and came up to Brougham Bridge. That is to say, I then and there felt the galvanic circuit of thought closed, and the sparks which fell from it were the fundamental equations between i, j, k, *exactly such* as I have used them ever since. I pulled out, on the spot, a pocketbook, which still exists, and made an entry, on which, *at the very moment*, I felt that it might be worth my while to expend the labour of at least ten (or it might be fifteen) years to come. But then it is fair to say that this was because I felt a *problem* to have been at the moment *solved*, an intellectual *want relieved*, which had haunted me for at least fifteen years before.

So struck was W.R. Hamilton with his discovery that he drew his pocketknife and into a stone of Broome Bridge in Dublin he carved the famous formulae with the symbols i, j, k:

$$i^2 = j^2 = k^2 = ijk = -1.$$

Thus, the Quaternions were born. What W.R. Hamilton had imagined was a number with one real component and three distinct imaginary components, all of the imaginary components squaring to -1. Unfortunately, the carvings no longer remain today. Nevertheless, his discovery was so significant that every year on October 16, the Mathematics Department of the National University of Ireland, Maynooth, holds a Hamilton Walk to Broome Bridge commemorating his discovery. This is the breakthrough concerning the designation of quaternions—from the Latin quaternio, meaning "set of four".

W.R. Hamilton presented quaternion mathematics at a series of lectures at the Royal Irish Academy. The lectures gave rise to a book, whose full title is: "Lectures on quaternions: containing a systematic statement of a new mathematical method, of which the principles were communicated in 1843 to the Royal Irish academy, and which has since formed the subject of successive courses of lectures, delivered in 1848 and subsequent years, in the halls of Trinity college, Dublin: with numerous illustrative diagrams, and with some geometrical and physical applications".

Leaving history behind us, the next portion of text intends to introduce the reader to the modern terminology and basic notations used to describe quaternions. The standard properties and the arithmetics of quaternions are presented. It will be shown that quaternions have a simple four-dimensional character, leading to a straightforward geometric description. When necessary, some important constructions are explained in full detail. And for the reader's convenience and sake of easy reference, from now on the main results exploited here are systematically stated and proved step-by-step. Nevertheless, a rigorous development has to be replaced by a logical one based upon suitable definitions in order to agree with the given notation. In doing so, this will allow us to fully concentrate on examples and exercises since our experience has shown that this methodology is useful for someone who is introduced to the subject for the first time.

W.R. Hamilton called the new numbers

$$p := a + bi + cj + dk \quad (a, b, c, d \in \mathbb{R})$$

real quaternions (or more informally, Hamilton numbers). The set of all real quaternions is often denoted by $\mathbb{H}$, in honour of its discoverer. We note that it is possible to consider a more general structure, which is of no particular interest for our objectives; known as the algebra of complex quaternions (or biquaternion algebra).

1.1 Basic Units

For the remainder of this section, 1, i, j, and k will denote both imaginary basis units of $\mathbb{H}$ and a basis for $\mathbb{R}^4$. For the sake of clarity, the Hamiltonian rule

$$i^2 = j^2 = k^2 = ijk = -1$$

may also be written as:

$$ij = k = -ji, \quad jk = i = -kj, \quad \text{and} \quad ki = j = -ik. \tag{1.1}$$

In particular, the elements i, j, k pairwise anticommute. When convenient, it can be very useful to use another representation of quaternions, which will be introduced in the following. Geometrically speaking, a quaternion which is initially given in the form $p = a + bi + cj + dk$ can be unequivocally associated to an ordered quadruple of real numbers (a, b, c, d). The starting point for this consideration is the identification of the basis elements of $\mathbb{R}^4$, $(1, 0, 0, 0)$, $(0, 1, 0, 0)$, $(0, 0, 1, 0)$ and $(0, 0, 0, 1)$, with the basis of $\mathbb{H}$, 1, i, j, and k, respectively. As a matter of fact, throughout the text we will often use the same symbol to represent a point in $\mathbb{R}^4$ as well as the corresponding quaternion.

1.2   Scalar and Vector Parts

The notations $p = a + ib + jc + kd$ and $p = a + bi + cj + dk$ are interchangeable. The real number $p_0 := a$ in p is called the scalar part of p and $\mathbf{p} := bi + cj + dk$ is the vector part of p. The scalar and vector parts of a real quaternion p are also often abbreviated as $\mathrm{Sc}(p)$ and $\mathrm{Vec}(p)$, respectively. For example, if $p = 1 + i + 4j + 3k$, then $p_0 = 1$ and $\mathbf{p} = i + 4j + 3k$. The real numbers are precisely those with zero vector part. In addition, if $p = \mathbf{p}$, then p is called a pure imaginary quaternion or, simply, pure quaternion.[1] For example, $p = j + k$ is a pure quaternion. If p_1 and p_2 are two quaternions, we generally do not define inequalities like $p_1 \leq p_2$ unless p_1 and p_2 are real numbers. The words positive and negative are not applied to quaternions either, and the use of these words implies that we are dealing again with real numbers.

1.3   Convention

Even though every quaternion p can be uniquely associated to a vector $(a, b, c, d)^T$ in $\mathbb{R}^4$ (hereby transposition shows that in general vectors should be written as column vectors), one can often define a vector to mean a pure quaternion. Ultimately, this convention can be reduced so that a vector is an element of the vector space $\mathbb{R}^3$. In other courses you have undoubtedly seen that the numbers in an ordered triple of real numbers can be interpreted as the components of a vector in $\mathbb{R}^3$. Thus, a (pure) quaternion p can also be viewed as a three-dimensional position vector, that is, a vector whose initial point is the origin and whose terminal point is (b, c, d). Even nowadays, Hamilton's notation i, j, and k for the basis elements in $\mathbb{R}^3$ is often used in engineering lectures. Thus a vector in the sense of Hamilton is representable in the form $\mathbf{p} := bi + cj + dk$.

---

[1] Hamilton called pure imaginary quaternions right quaternions and real numbers scalar quaternions.

Because of the previous correspondence between a pure quaternion $\mathbf{p} = bi + cj + dk$ and one and only one point (b, c, d) in $\mathbb{R}^3$, we shall from now on use the terms pure quaternion and point in $\mathbb{R}^3$ interchangeably.

1.4 Equality

Two quaternions $p = a_1 + b_1 i + c_1 j + d_1 k$ and $q = a_2 + b_2 i + c_2 j + d_2 k$ are equal, $p = q$, when the individual coordinates are equal: $a_1 = a_2, b_1 = b_2, c_1 = c_2$ and $d_1 = d_2$. In terms of the scalar and vector parts, the last statement means that $p = q$ if $p_0 = q_0$ and $\mathbf{p} = \mathbf{q}$.

1.5 Arithmetic Operations

Quaternions can be added, subtracted and multiplied. If $p = a_1 + b_1 i + c_1 j + d_1 k$, $q = a_2 + b_2 i + c_2 j + d_2 k$ and λ is a real number, these operations are defined as follows:

(i) Addition
$$p + q := (a_1 + a_2) + (b_1 + b_2)i + (c_1 + c_2)j + (d_1 + d_2)k;$$
(ii) Subtraction
$$p - q := (a_1 - a_2) + (b_1 - b_2)i + (c_1 - c_2)j + (d_1 - d_2)k;$$
(iii) Real multiplication
$$\lambda p := (\lambda a_1) + (\lambda b_1)i + (\lambda c_1)j + (\lambda d_1)k;$$
(iv) Quaternion multiplication
$$pq := (a_1 a_2 - b_1 b_2 - c_1 c_2 - d_1 d_2) + (a_1 b_2 + b_1 a_2 + c_1 d_2 - d_1 c_2)i$$
$$+ (a_1 c_2 - b_1 d_2 + c_1 a_2 + d_1 b_2)j + (a_1 d_2 + b_1 c_2 - c_1 b_2 + d_1 a_2)k.$$

There are several ways to assemble Rule (iv). For the considerations to follow we use the (left-)distributivity of multiplication over addition; then the quaternion multiplication of p and q becomes

$$pq := (a_1 + b_1 i + c_1 j + d_1 k)(a_2 + b_2 i + c_2 j + d_2 k)$$
$$= a_1 a_2 + a_1 b_2 i + a_1 c_2 j + a_1 d_2 k + b_1 a_2 i + b_1 b_2 i^2 + b_1 c_2 ij + b_1 d_2 ik$$
$$+ c_1 a_2 j + c_1 b_2 ji + c_1 c_2 j^2 + c_1 d_2 jk + d_1 a_2 k + d_1 b_2 ki + d_1 c_2 kj$$
$$+ d_1 d_2 k^2.$$

Notice that we must be careful simplifying this last step because although multiplication is commutative for real numbers, this is no longer true for imaginary elements. Using then the basic properties of quaternions and identities (1.1), we can rewrite the multiplication again:

$$pq = a_1 a_2 + a_1 b_2 i + a_1 c_2 j + a_1 d_2 k + b_1 a_2 i - b_1 b_2 + b_1 c_2 k - b_1 d_2 j$$
$$+ c_1 a_2 j - c_1 b_2 k - c_1 c_2 + c_1 d_2 i + d_1 a_2 k + d_1 b_2 j - d_1 c_2 i - d_1 d_2.$$

Regrouping the terms according to the imaginary units, we finally get Rule (iv). A detailed verification of the remaining rules is left to the reader.

If we take the expressions above as the definitions of addition and multiplication, it is a simple matter to verify that the familiar associativity, and distributivity laws of multiplication over addition hold for quaternions:

(v) Commutativity law of addition
$$p + q = q + p \quad \text{for all } p, q \in \mathbb{H};$$
(vi) Associativity law of addition
$$p + (q + r) = (p + q) + r \quad \text{for all } p, q, r \in \mathbb{H};$$
(vii) Distributivity law of multiplication over addition
$$p(q + r) = pq + pr \quad \text{and} \quad (q + r)p = qp + rp \quad \text{for all } p, q, r \in \mathbb{H};$$
(viii) Associativity law of multiplication
$$(pq)r = p(qr) \quad \text{for all } p, q, r \in \mathbb{H}.$$

Exercise 1.1. *Prove the previous properties.*

All that is needed to understand the definitions of addition, subtraction, and multiplication is:

(i) *To add (subtract) two quaternions, simply add (subtract) the corresponding scalar and vector parts, or simply use the usual sum (subtraction) of two vectors in $\mathbb{R}^4$;*

(ii) *To multiply two real quaternions, use the distributivity law of multiplication over addition or use the following scheme:*

	1	i	j	k
1	1	i	j	k
i	i	-1	k	$-j$
j	j	$-k$	-1	i
k	k	j	$-i$	-1

The previous presentation is called Cayley table, and it was developed by Arthur Cayley, an English mathematician (1821–1895). Also, the following diagram is often very useful:

$$
\begin{array}{ccc}
 & k & \\
\swarrow & & \nwarrow \\
i & \longrightarrow & j
\end{array}
$$

Although addition is commutative, the same does not hold for multiplication: for example, $ij = k$, while $ji = -k$. Therefore, in general, pq is not equal to qp.

Example. If $p = 3 + i - j$ and $q = j - 2k$, compute (a) $p + q$; (b) $3p - q$; (c) pq; (d) qp.

Solution. (a) By adding scalar and vector parts, the sum of the two quaternions p and q is

$$p + q = (3 + 0) + (1 + 0)i + (-1 + 1)j + (0 - 2)k = 3 + i - 2k.$$

(b) In the same way, we have

$$3p - q = (9 + 0) + (3 + 0)i + (-3 - 1)j + (0 + 2)k = 9 + 3i - 4j + 2k.$$

(c) and (d) Using Cayley's table, direct computations show that

$$pq = 3j - 6k + ij - 2ik - j^2 + 2jk = 1 + 2i + 5j - 5k$$

and

$$qp = 1 - 2i + j - 7k \neq pq.$$

Exercise 1.2. *Let* p, q *and* r *be arbitrary elements in* $\mathbb{H}$. *Prove that* $\mathrm{Sc}(pqr) = \mathrm{Sc}(qrp) = \mathrm{Sc}(rpq)$.

Exercise 1.3. *If* $p = 1 + \lambda i - j + k$ *and* $q = i - \lambda j - \frac{1}{\lambda}k$ $(\lambda \in \mathbb{R} \setminus \{0\})$, *find* λ *such that* pq *and* pqp *are pure quaternions.*

Solution. $\lambda = \pm \frac{\sqrt{2}}{2}$.

The definition of division deserves further elaboration, and so we will discuss this point in more detail shortly.

1.6 Special Quaternions

It is clear that the quaternion $0 + 0i + 0j + 0k =: 0_{\mathbb{H}}$ is the neutral element of addition, known as additive identity quaternion, and the quaternion $1 + 0i + 0j + 0k =: 1_{\mathbb{H}}$ is the multiplicative identity quaternion. They are such that any quaternion added or multiplied by them remains unchanged. If we take a previous expression as the definition of addition, it is a simple matter to verify that the inverse element of p with respect to addition is $-p$. The inverse with respect to multiplication will be further detailed below. In summary, the real quaternions form a noncommutative division algebra, the skew field[2] of the real quaternions $\mathbb{H}$. The quaternions remain the simplest algebra after the real and complex numbers. Indeed, the real numbers, the complex numbers, and the quaternions are the only associative division algebras,

---

[2] A skew field is an algebraic structure that satisfies all the properties of a field except for commutativity.

as was proved by Georg Frobenius (1849–1917); and amongst these the quaternions are the most general. Since any complex number $a + bi$ can be written as $a + bi + 0j + 0k$, we see that the set of complex numbers can be regarded as a subset of $\mathbb{H}$. We could also have chosen j or k to play the role of the imaginary unit $i = \sqrt{-1}$ in classical complex analysis.

The assumption $ij = ji$ is incompatible with the request that $\mathbb{H}$ is a skew symmetric field. Even though it may not be so common, it is possible to consider such a structure, which is of independent interest and is known as the commutative ring of bicomplex numbers.

1.7   Decomposition of Quaternions

Quaternions can be conveniently decomposed in two ways: $p = a + ib + cj + idj$ or $p = p_+ + p_- = \frac{1}{2}(p + ipj) + \frac{1}{2}(p - ipj)$. In this connection, for the second decomposition it holds that

$$p_\pm = \left(a \pm d + i(b \mp c)\right)\frac{(1 \pm k)}{2} = \frac{(1 \pm k)}{2}\left(a \pm d + j(c \mp b)\right).$$

(See the list of exercises on pages 33 and 34 for some applications and further properties using this decomposition.)

Exercise 1.4. *Let $p = -1 + i + 7j - 3k$ and $q = -i - j + 2k$. Find p_+, q_+, p_- and q_-.*

Solution. $p_+ = -2 - 3i + 3j - 2k,\ q_+ = 1 + k,\ p_- = 1 + 4i + 4j - k$, and $q_- = -1 - i - j + k$.

Exercise 1.5. *Find $p = a + bi + cj + dk \in \mathbb{H}$ such that $p_+ = 0_{\mathbb{H}}$.*

Solution. $p = a + bi + bj - ak\ (a, b \in \mathbb{R})$.

Exercise 1.6. *Find $p = a + bi + cj + dk \in \mathbb{H}$ such that $p_- = 0_{\mathbb{H}}$.*

Solution. $p = a + bi - bj + ak\ (a, b \in \mathbb{R})$.

Exercise 1.7. *Let $p \in \mathbb{H}$. Prove that $p_+ = p_-$ if and only if $p \equiv 0_{\mathbb{H}}$.*

Exercise 1.8. *Let $p = a + bi + cj + dk \in \mathbb{H}$. Compute p_+p_-.*

Solution. $\frac{1}{2}\left[a^2 - d^2 + (ab + cd - ac - bd)i + (ab + ac + bd + cd)j + (b^2 - c^2)k\right]$.

1.8   Roots

In the sequel, let q be a real quaternion and $n \in \mathbb{N}$. As a brief preview of the material in Chap. 5, a quaternion p is called (an) n-th root of q if $p^n = q$. The noncommutativity of multiplication has some unexpected consequences, for instance, polynomial equations over the quaternions can have more distinct solutions than the degree of the polynomial. For example, the equation $p^2 + 1_{\mathbb{H}} = 0_{\mathbb{H}}$ has an infinite number of pure quaternion solutions $p = bi + cj + dk$ with $b^2 + c^2 + d^2 = 1$. To see this, let $p = a + bi + cj + dk$ be a quaternion, and assume that its square is $-1_{\mathbb{H}}$. In terms of a, b, c, and d, this leads to the following equations: $a^2 - b^2 - c^2 - d^2 = -1$, $2ab = 0$, $2ac = 0$, and $2ad = 0$. To satisfy the last three equations, either $a = 0$ or b, c, and d are all 0. The latter is impossible because a is a real number and the first equation would imply that $a^2 = -1$; therefore $a = 0$ and $b^2 + c^2 + d^2 = 1$. As the above discussion shows, a quaternion squares to $-1_{\mathbb{H}}$ if and only if it is a vector in $\mathbb{R}^3$ with norm 1. Therefore, these solutions form a two-dimensional sphere centered at zero in the three-dimensional subspace of pure imaginary quaternions. This sphere intersects the complex plane at the two poles i and $-i$, respectively.[3]

1.9   Quaternion Conjugation

The quaternion $\overline{p} := p_0 - \mathbf{p} = a - bi - cj - dk$ associated with the quaternion $p := a + bi + cj + dk$, defined by reversing the sign of the vector part of p, is called the conjugate quaternion of p. For example, if $p = 1 - 2i + 3j + k$, then $\overline{p} = 1 + 2i - 3j - k$; if $p = -1 - j + k$, then $\overline{p} = -1 + j - k$. Notice that replacing i by $-i$, j by $-j$, and k by $-k$ sends a vector in $\mathbb{R}^3$ to its additive inverse, so the additive inverse of a vector in $\mathbb{R}^3$ is the same as its conjugate as a pure quaternion. For this reason, conjugation is sometimes called the spatial inverse. The quaternion conjugation can also be useful to extract the scalar and vector parts of p. In this sense, the scalar part of p is $(p + \overline{p})/2$, and the vector part of p is $(p - \overline{p})/2$. For all $p, q \in \mathbb{H}$ and $\lambda \in \mathbb{R}$, the mapping $p \mapsto \overline{p}$ has the following properties:

(i) $\overline{p \pm q} = \overline{p} \pm \overline{q}$;   (ii) $\overline{\overline{p}} = p$;

(iii) $\overline{\lambda p} = \lambda \overline{p}$;   (iv) $\overline{pq} = \overline{q}\,\overline{p}$;

(v) $p \in \mathbb{R} \Leftrightarrow p = \overline{p}$;   (vi) p is a pure quaternion $\Leftrightarrow p = -\overline{p}$.

The definitions of addition and multiplication show that the sum and the product of a real quaternion $p = a + bi + cj + dk$ with its conjugate $\overline{p}$ are real numbers: $p + \overline{p} = (a + bi + cj + dk) + (a - bi - cj - dk) = 2a$, and $p\overline{p} = (a + bi + cj + dk)(a - bi - cj - dk) = a^2 + b^2 + c^2 + d^2$. The difference of a real quaternion p with its conjugate $\overline{p}$ is a pure quaternion: $p - \overline{p} = (a + bi + cj + dk) - (a - bi - cj - dk) = 2bi + 2cj + 2dk$. Properties (i) and (iv) can be extended to three or more quaternions. For example

---

[3]For all real angles θ and ϕ, it is easy to check that $p = \cos\theta \cos\phi\, i + \cos\theta \sin\phi\, j + \sin\theta k$ is a square root of $-1_{\mathbb{H}}$. This gives the set of all the points on the unit sphere in $\mathbb{R}^3$ and shows that the set of "quaternionic square roots of minus one" has infinitely many square roots.

$\overline{p_1 \pm p_2 \pm \cdots \pm p_n} = \overline{p_1} \pm \overline{p_2} \pm \cdots \pm \overline{p_n}$. Similarly, $\overline{p_1 p_2 p_3} = \overline{p_3}\,\overline{p_2}\,\overline{p_1}$, or, in general $\overline{p_1 p_2 \cdots p_{n-1} p_n} = \overline{p_n}\,\overline{p_{n-1}} \cdots \overline{p_2}\,\overline{p_1}$. In summary, the conjugate of a sum (difference) of quaternions is the sum (difference) of the conjugates of the summands. The conjugate of a product of quaternions is the product of the conjugates of each factor in the reverse order.

Exercise 1.9. *Solve the following equations for $p = a + bi + cj + dk$:* (a) $p^2 = -j$; (b) $\overline{p}^2 = i - k$;   (c) $p - \frac{1}{7}\overline{p} - k = 0_{\mathbb{H}}$;   (d) $\frac{1}{3} p \overline{p} - ijk = 0_{\mathbb{H}}$;   (e) $(p - \overline{p})p = 1 + i + j + k$.

Solution. (a) $p = \pm \frac{\sqrt{2}}{2} \mp \frac{\sqrt{2}}{2} j$;   (b) $p = \mp \frac{2^{3/4}}{2} \pm \frac{2^{1/4}}{2} i \mp \frac{2^{1/4}}{2} k$;   (c) $p = \frac{7}{8} k$; (d) No solutions;   (e) No solutions.

1.10 Quaternion Modulus and Quaternion Inverse

Associated with every quaternion is a nonnegative real number, which we now define. The modulus or absolute value[4] of a real quaternion $p = a + bi + cj + dk$ is denoted by $|p|$ and is identical to the notion of Euclidean length in $\mathbb{R}^4$, so that it is clearly the distance from the origin to the point (a, b, c, d). More precisely, we define the modulus of p by the expression $|p| := \sqrt{a^2 + b^2 + c^2 + d^2}$. If $|p| = 1$, the quaternion is called a unit quaternion. Using conjugation and the modulus makes it possible to define the multiplicative inverse of a quaternion p in such a way that the product of a quaternion with its inverse (in either order) is $1_{\mathbb{H}}$. In symbols, for $p \neq 0_{\mathbb{H}}$ there exists one and only one nonzero quaternion p^{-1} such that $pp^{-1} = p^{-1}p = 1_{\mathbb{H}}$. The inverse of a unit quaternion is its conjugate. In particular, the following properties hold:

(i) $p\overline{p} = \overline{p}p = |p|^2$;

(ii) $p^{-1} = \frac{\overline{p}}{|p|^2}$, $p \neq 0_{\mathbb{H}}$;

(iii) $|pq| = |p||q|$;

(iv) $|\overline{p}| = |-p| = |p|$;

(v) $(pq)^{-1} = q^{-1}p^{-1}$; $pq \neq 0_{\mathbb{H}}$.

Statement (ii) shows that the quaternions are indeed a skew field, that means, every nonzero element has a multiplicative inverse. Statement (iii) shows that the product pq is a quaternion and so a vector in $\mathbb{R}^4$. The equality $|pq| = |p||q|$ hints that the length of the quaternion pq is exactly the product of the lengths of the individual quaternions p and q. For simplicity we only prove Statements (ii) and (v). A detailed verification of the remaining properties is left to the reader. We can easily verify Statement (ii) by performing the multiplication

---

[4]Hamilton called this quantity the tensor of p, but nowadays this concept conflicts with modern usage.

$$pp^{-1} = \frac{p\overline{p}}{|p|^2} = \frac{|p|^2}{|p|^2} = 1_{\mathbb{H}}.$$

Similarly, $p^{-1}p = \frac{\overline{p}p}{|p|^2} = \frac{|p|^2}{|p|^2} = 1_{\mathbb{H}}$. Therefore, we have $(pq)(pq)^{-1} = 1_{\mathbb{H}}$ and, analogously, $(pq)^{-1}(pq) = 1_{\mathbb{H}}$. For Statement (v) one has $p^{-1}pq(pq)^{-1} = p^{-1}$. Hence $q^{-1}q(pq)^{-1} = q^{-1}p^{-1}$. That is, $(pq)^{-1} = q^{-1}p^{-1}$.

Exercise 1.10. *Let p be a real quaternion. Show that:* (a) $|p| = |-p|$; *(b)* $|p| = |\overline{p}|$.

Exercise 1.11. *Let p be a real quaternion. Prove that $p^2 - 2p_0 p + |p|^2 = 0_{\mathbb{H}}$.*

Exercise 1.12. *What can be said about a real quaternion p if $p^2 = (\overline{p})^2$? Making use of this fact, under what circumstances does $p^2 = (p^{-1})^2$ hold?*

Exercise 1.13. *Let p and q be two real quaternions. Show that*
(a) $|p + q|^2 + |p - q|^2 = 2\left(|p|^2 + |q|^2\right)$;
(b) $|p + q|^2 = |p|^2 + 2\mathrm{Sc}(p\overline{q}) + |q|^2$;
(c) $\left(|p| + |q|\right)^2 = |p|^2 + 2|p\overline{q}| + |q|^2$.

Exercise 1.14. *Given a unit quaternion q and a pure quaternion p, define $T_q(p) = q\overline{p}q^{-1}$.*
(a) *Prove that $T_p(p)$ is another pure quaternion;*
(b) *Show that $|T_q(p)| = |p|$;*
(c) *Let $\lambda \in \mathbb{R}$ and let p, q be real quaternions. Prove that T_q is linear: $T_q(\lambda p + q) = \lambda T_q(p) + T_q(q)$.*

Exercise 1.15. *Given a pure quaternion p, prove that $p^{2n} = (-1)^n|p|^{2n}$ for all $n \in \mathbb{N}$. Making use of this fact prove that*

$$\sum_{n=0}^{\infty} \frac{p^n}{n!} = \cos|p| + \frac{p}{|p|}\sin|p|.$$

Exercise 1.16. *Let $a, b \in \mathbb{R}$ such that $a^2 - 4b < 0$. Use the previous exercise to show that for every given pure quaternion p we have $p^{2n} - ap^n + b > 0$ when n is even, and $p^{2n} - ap^n + b < 0$ when n is odd.*

1.11 Quaternion Quotient

With regard to the concept of quaternion inverse discussed above and due to the noncommutativity of multiplication, we assume familiarity with the fact that the quotient of two quaternions p and q may be interpreted in two different ways.

That is, their quotient can be either pq^{-1} or $q^{-1}p$. In this sense, the notation $\frac{p}{q}$ is ambiguous because it does not specify whether q divides on the left or the right. Namely, $pq^{-1} = \frac{p\bar{q}}{|q|^2}$, and $q^{-1}p = \frac{\bar{q}p}{|q|^2}$. For conciseness the first is called right quotient and the second left quotient. It is significant to note that when p and q are pure quaternions then the right and left quotients, pq^{-1} and $q^{-1}p$, are quaternion conjugates. Take for example $p = i + j$ and $q = k$. The right and left quotients of p and q are, respectively, $pq^{-1} = -i + j$ and $q^{-1}p = i - j$. We further note that $\overline{pq^{-1}} = i - j = q^{-1}p$.

Exercise 1.17. *Compute the right and left quotients, pq^{-1} and $q^{-1}p$, of* (a) $p = j - 3k, q = 1 + i;$ (b) $p = i + j + k, q = \frac{1}{11}(-j + 9k)$.

Solution. (a) $pq^{-1} = 2j - k, q^{-1}p = -(j + 2k);$ (b) $pq^{-1} = \frac{44}{41} - \frac{55}{41}i + \frac{99}{82}j + \frac{11}{82}k, q^{-1}p = \frac{44}{41} + \frac{55}{41}i - \frac{99}{82}j - \frac{11}{82}k.$

1.12 Triangle Inequalities

Although the concept of order in the real number system does not carry over to the quaternion number system, since $|p|$ is a real number we can compare the moduli of two real quaternions. For example, if $p = 1 - 2i$ and $q = i - j + k$, then $|p| = \sqrt{5}$ and $|q| = \sqrt{3}$ and, consequently, $|p| > |q|$. We record now some useful inequalities. Let p and q be quaternions, then for the modulus we have

(i) $-|p| \leq p_0 \leq |p|$ and $-|p| \leq |\mathbf{p}| \leq |p|$, that is $|p_0| \leq |p|$ and $|\mathbf{p}| \leq |p|$;

(ii) $|p \pm q| \leq |p| + |q|$;   (iii) $||p| - |q|| \leq |p \pm q|$.

For simplicity we only sketch the proof of $||p| - |q|| \leq |p + q|$. A detailed verification of the remaining inequalities is left to the reader. From the identity $p = p + q + (-q)$, Statement (ii) gives $|p| \leq |p + q| + |-q|$. Since $|q| = |-q|$, solving the last inequality for $|p + q|$ yields the relation $|p + q| \geq |p| - |q|$. Now because $p + q = q + p$, the last inequality can be rewritten as $|p + q| = |q + p| \geq |q| - |p| = -(|p| - |q|)$, and so combined with the last relation implies $|p + q| \geq ||p| - |q||$. The triangle inequality: $|p + q| \leq |p| + |q|$ indicates that the length of the quaternion $p + q$ cannot exceed the sum of the lengths of the individual quaternions p and q. In addition, we note that this inequality extends to any finite sum of quaternions, so that $|p_1 + p_2 + \cdots + p_n| \leq |p_1| + |p_2| + \cdots + |p_n|$.

Example. Consider two quaternions p and q. Under what circumstances does $|p + q| = |p| + |q|$ hold?

Solution. Let $p = p_0 + p_1 i + p_2 j + p_3 k$ and $q = q_0 + q_1 i + q_2 j + q_3 k$. It is readily verified that $|p + q| = |p| + |q|$ is equivalent to $|p + q|^2 = |p|^2 + |q|^2 + 2|p||q|$. The last equality is only possible if $p_l = \lambda q_l$ $(l = 0, 1, 2, 3)$ for some $\lambda > 0$. Hence $p \equiv \lambda q$ $(\lambda > 0)$.

Exercise 1.18. *Let $p = a + bi + cj + dk$ such that $|p| = 2$. Find $q \in \mathbb{R}$ such that*
$$\left| \frac{1}{p^4 - qp^2 + 3} \right| \leq \tfrac{1}{3}.$$

Solution. $q \in [-4, 4]$.

Exercise 1.19. *Let p be a real quaternion. Find a lower bound for the modulus of $p^3 - p + 11$ if $|p| \leq 1$.*

Exercise 1.20. *Consider two quaternions p and q of absolute value < 1. Show that $|p - q| < |1 - \overline{p}q|$.*

By the early 1880s, Josiah Willard Gibbs (1839–1903) introduced the dot product and cross product of vectors. Gibbs recognized that a pure quaternion could be interpreted as a vector, which allowed him to build the basis of a vectorial system. Historically, vector analysis developed by Gibbs and his collaborators is seen as the origin of practical applications of physics and engineering, which were missing in the more complicated quaternion language.

1.13 Quaternion Dot Product

The dot product of two real quaternions p and q is a real number, defined as the sum of the quantity of each element of p multiplied by each element of q, that is

$$p \cdot q := p_0 q_0 + \mathbf{p} \cdot \mathbf{q}.$$

Take for example $p = 1 + i + j + k$ and $q = 1 - 2i + 3j + 5k$. A simple calculation shows that $p \cdot q = 1 - 2 + 3 + 5 = 7$. The quaternion dot product is also referred to as the Euclidean inner product for vectors in $\mathbb{R}^4$, as it returns a scalar quantity. If $p \cdot q = 0$, then p and q are orthogonal to each other. This product can be used to extract any component from a given quaternion. For instance, the i component can be extracted from $p = a + bi + cj + dk$ by taking $p \cdot i = b$. Also, it holds that

$$p \cdot q = \mathrm{Sc}(p\overline{q}) = \mathrm{Sc}(q\overline{p})$$

$$= \frac{1}{2}(\overline{p}q + \overline{q}p) = \frac{1}{2}(p\overline{q} + q\overline{p}).$$

Exercise 1.21. *Let p, q and r be arbitrary quaternions. Prove that the quaternion dot product is both left and right distributive, i.e. $p \cdot (q + r) = p \cdot q + p \cdot r$ and $(p + q) \cdot r = p \cdot r + q \cdot r$. Is the quaternion dot product commutative? Justify your answer.*

Exercise 1.22. *Let p be a quaternion element. Prove that $|p|^2 = p_0^2 + \mathbf{p} \cdot \mathbf{p}$.*

Exercise 1.23. *Let p and q be two quaternions. Prove that* $\mathbf{p} \cdot \mathbf{q} = -\mathrm{Sc}(\mathbf{pq}) = -\frac{1}{2}(\mathbf{pq} + \mathbf{qp})$.

Exercise 1.24. *Let p, q and r be quaternions. Prove that* $\mathrm{Sc}(pqr) = \overline{p} \cdot (qr)$.

Exercise 1.25. *Let p and q be two quaternions. Prove the polarization identity* $p \cdot q = \frac{1}{4}\left(|p+q|^2 - |p-q|^2\right)$.

Exercise 1.26. *Let p and q be two quaternions. Prove the well-known Schwarz inequality* $|p \cdot q| \leq |p||q|$.

Exercise 1.27. *Let $p \in \mathbb{H}$. Compute $p_+ \cdot p_-$.*

Solution. 0.

In the case of a quaternion consisting only of a vector part, W.R. Hamilton defined close analogues of the modern cross, outer and even products.

1.14 Quaternion Cross Product

In the sequel, consider two pure quaternions $\mathbf{p} = p_1 i + p_2 j + p_3 k$ and $\mathbf{q} = q_1 i + q_2 j + q_3 k$. The quaternion cross product is also known as the odd product. It is equivalent to the vector cross product, and returns a vector quantity only:

$$\mathbf{p} \times \mathbf{q} := \mathrm{Vec}(\mathbf{pq}).$$

The definition of the cross product can also be represented by the determinant of a formal 3×3 matrix:

$$\det \begin{pmatrix} i & j & k \\ p_1 & p_2 & p_3 \\ q_1 & q_2 & q_3 \end{pmatrix} = (p_2 q_3 - q_2 p_3)i + (q_1 p_3 - p_1 q_3)j \\ + (p_1 q_2 - q_1 p_2)k.$$

The reader might know that this determinant can be computed using Sarrus' rule or cofactor expansion. It has the following properties:

(i) Anticommutativity

  $\mathbf{p} \times \mathbf{q} = -\mathbf{q} \times \mathbf{p}$ for pure quaternions $\mathbf{p}, \mathbf{q}$;

(ii) Distributivity law of the cross product over addition

  $\mathbf{p} \times (\mathbf{q} + \mathbf{r}) = (\mathbf{p} \times \mathbf{q}) + (\mathbf{p} \times \mathbf{r})$ for pure quaternions $\mathbf{p}, \mathbf{q}, \mathbf{r}$;

(iii) Multiplication by scalars

  $(\lambda \mathbf{p}) \times \mathbf{q} = \mathbf{p} \times (\lambda \mathbf{q}) = \lambda(\mathbf{p} \times \mathbf{q})$ for pure quaternions $\mathbf{p}, \mathbf{q}$ and $\lambda \in \mathbb{R}$;

(iv) Jacobi identity

  $\mathbf{p} \times (\mathbf{q} \times \mathbf{r}) + \mathbf{q} \times (\mathbf{r} \times \mathbf{p}) + \mathbf{r} \times (\mathbf{p} \times \mathbf{q}) = 0_{\mathbb{H}}$ for pure quaternions $\mathbf{p}, \mathbf{q}, \mathbf{r}$.

Example. Find $\mathbf{p} \times \mathbf{q}$, where $\mathbf{p} = 3i + j - k$ and $\mathbf{q} = i - j - k$.

Solution. The cross product of $\mathbf{p}$ and $\mathbf{q}$ is

$$\mathbf{p} \times \mathbf{q} = \det \begin{pmatrix} i & j & k \\ 3 & 1 & -1 \\ 1 & -1 & -1 \end{pmatrix}.$$

Exercise 1.28. *Compute:* (a) $2i \times (i + j)$;   (b) $(i + j) \times (i - j)$; (c) $((i + j) \times i) \times j$;   (d) $(i + j) \times (i \times j)$.

Solution. (a) $2k$;   (b) $-2k$;   (c) i;   (d) $i - j$.

Exercise 1.29. *Prove that the standard basis vectors i, j, and k satisfy the following equalities:* (a) $i \times j = k$;   (b) $j \times k = i$;   (c) $k \times i = j$.

Exercise 1.30. *Let $\mathbf{p}$ and $\mathbf{q}$ be pure quaternions. Prove that $\mathbf{p} \times \mathbf{q} = \frac{1}{2}(\mathbf{pq} - \mathbf{qp})$.*

Exercise 1.31. *Let $\mathbf{p}$ and $\mathbf{q}$ be pure quaternions. Prove that $\mathbf{p} \times \mathbf{q} = \frac{1}{2}(\mathbf{pq} - \overline{\mathbf{pq}})$.*

Exercise 1.32. *Let $\mathbf{p}$ and $\mathbf{q}$ be pure quaternions. Show that $\mathbf{pq} = -\mathbf{p} \cdot \mathbf{q} + \mathbf{p} \times \mathbf{q}$.*

Exercise 1.33. *Let $\mathbf{p}$ and $\mathbf{q}$ be pure quaternions. Show that $\overline{\mathbf{pq}} = \overline{\mathbf{q}} \times \overline{\mathbf{p}}$.*

Exercise 1.34. *Let $p = a + bi + cj + dk \in \mathbb{H}$. Find conditions for the parameters a, b, c, d so that $p_\pm$ are pure quaternions. For those values of a, b, c, d, compute $\mathbf{p}_+ \times \mathbf{p}_-$.*

Solution. $p_\pm$ are both pure quaternions if and only if $a = d = 0$. Hence $\mathbf{p}_+ \times \mathbf{p}_- = \frac{(b^2 - c^2)}{2}k$.

According to the above discussion, a more precise result holds true:

$$
\begin{aligned}
pq &= (p_0 + \mathbf{p})(q_0 + \mathbf{q}) \\
&= p_0 q_0 - (p_1 q_1 + p_2 q_2 + p_3 q_3) + (p_0 q_1 + q_0 p_1 + p_2 q_3 - p_3 q_2)i \\
&\quad + (p_0 q_2 + q_0 p_2 + p_3 q_1 - p_1 q_3)j + (p_0 q_3 + q_0 p_3 + p_1 q_2 - p_2 q_1)k \\
&= (p_0 q_0 - \mathbf{p} \cdot \mathbf{q}) + p_0(q_1 i + q_2 j + q_3 k) + q_0(p_1 i + p_2 j + p_3 k) \\
&\quad + (p_2 q_3 - p_3 q_2)i + (p_3 q_1 - p_1 q_3)j + (p_1 q_2 - p_2 q_1)k \\
&= (p_0 q_0 - \mathbf{p} \cdot \mathbf{q}) + p_0 \mathbf{q} + \mathbf{p} q_0 + \mathbf{p} \times \mathbf{q}.
\end{aligned}
\tag{1.2}
$$

This evidences that the noncommutativity of multiplication comes from the multiplication of pure quaternions. In order to tie these matters up, we notice that two quaternions commute if and only if their vector parts are coplanar or collinear. In particular,

$$p^2 = (p_0^2 - \mathbf{p} \cdot \mathbf{p}) + 2p_0\mathbf{p},$$

since $\mathbf{p} \times \mathbf{p} = 0_{\mathbb{H}}$.

1.15   Mixed Product

For three pure quaternions $\mathbf{p}$, $\mathbf{q}$, and $\mathbf{r}$,

$$(\mathbf{p}, \mathbf{q}, \mathbf{r}) := \mathbf{p} \cdot (\mathbf{q} \times \mathbf{r})$$

is called their mixed product. The mixed product has the following properties:

(i) It defines a trilinear vector form, i.e. it is $\mathbb{R}$-homogeneous and distributive in each component. For example,

$$(\mathbf{p}, s\mathbf{q}_1 + t\mathbf{q}_2, \mathbf{r}) = s(\mathbf{p}, \mathbf{q}_1, \mathbf{r}) + t(\mathbf{p}, \mathbf{q}_2, \mathbf{r}), \quad s, t \in \mathbb{R};$$

(ii) It does not change if its factors are circularly permuted, but changes sign if they are switched, so that

$$(\mathbf{p}, \mathbf{q}, \mathbf{r}) = (\mathbf{q}, \mathbf{r}, \mathbf{p}) = (\mathbf{r}, \mathbf{p}, \mathbf{q}) = -(\mathbf{q}, \mathbf{p}, \mathbf{r}) = -(\mathbf{p}, \mathbf{r}, \mathbf{q}) = -(\mathbf{r}, \mathbf{q}, \mathbf{p});$$

(iii) The absolute value of the mixed product represents the oriented volume of the parallelepiped spanned by the three pure quaternions that meet in the same vertex. It follows that

$$(\mathbf{p}, \mathbf{q}, \mathbf{r}) = |\mathbf{p}||\mathbf{q}||\mathbf{r}| \sin \angle(\mathbf{p}, \mathbf{q}) \cos \angle(\mathbf{p} \times \mathbf{q}, \mathbf{r}).$$

Therefore, the mixed product vanishes if and only if the three pure quaternions are collinear;

(iv) The determinant representation holds:

$$(\mathbf{p}, \mathbf{q}, \mathbf{r}) = \det \begin{pmatrix} p_1 & p_2 & p_3 \\ q_1 & q_2 & q_3 \\ r_1 & r_2 & r_3 \end{pmatrix}$$

$$= p_1(q_2 r_3 - q_3 r_2) + p_2(q_3 r_1 - q_1 r_3) + p_3(q_1 r_2 - q_2 r_1).$$

For simplicity, we only prove Property (iv). The verification of the remaining properties is left to the reader. By definition, we have that

$$(\mathbf{p}, \mathbf{q}, \mathbf{r}) = (\mathbf{r}, \mathbf{p}, \mathbf{q}) = (\mathbf{p} \times \mathbf{q}) \cdot \mathbf{r}$$

$$= \left[\left(\begin{matrix} p_2 & p_3 \\ q_2 & q_3 \end{matrix} \right) i + \det \left(\begin{matrix} p_3 & p_1 \\ q_3 & q_1 \end{matrix} \right) j + \det \left(\begin{matrix} p_1 & p_2 \\ q_1 & q_2 \end{matrix} \right) k \right] \cdot (r_1 i + r_2 j + r_3 k)$$

$$= r_1 \det \left(\begin{matrix} p_2 & p_3 \\ q_2 & q_3 \end{matrix} \right) + r_2 \det \left(\begin{matrix} p_3 & p_1 \\ q_3 & q_1 \end{matrix} \right) + r_3 \det \left(\begin{matrix} p_1 & p_2 \\ q_1 & q_2 \end{matrix} \right).$$

Exercise 1.35. *Let* $\mathbf{p} = i + j$, $\mathbf{q} = 2i + 2j$, *and* $\mathbf{r} = k$. *Find* $(\mathbf{p}, \mathbf{q}, \mathbf{r})$.

Solution. 0.

1.16   Development Formula

Let $\mathbf{p}$, $\mathbf{q}$ and $\mathbf{r}$ be three pure quaternions. Then

$$\mathbf{p} \times (\mathbf{q} \times \mathbf{r}) = (\mathbf{p} \cdot \mathbf{r})\mathbf{q} - (\mathbf{p} \cdot \mathbf{q})\mathbf{r}. \tag{1.3}$$

Using the equality

$$2(\mathbf{p} \cdot \mathbf{r})\mathbf{q} = (\mathbf{p} \cdot \mathbf{r})\mathbf{q} + \mathbf{q}(\mathbf{p} \cdot \mathbf{r})$$

for each term from the definition of the scalar and vector products, we have

$$4[(\mathbf{p} \cdot \mathbf{r})\mathbf{q} - (\mathbf{p} \cdot \mathbf{q})\mathbf{r}] = (\mathbf{pr} + \mathbf{rp})\mathbf{q} + \mathbf{q}(\mathbf{pr} + \mathbf{rp}) - (\mathbf{pq} + \mathbf{qp})\mathbf{r}$$

$$- \mathbf{r}(\mathbf{pq} + \mathbf{qp})$$

$$= \mathbf{p}(\mathbf{rq} - \mathbf{qr}) + (\mathbf{qr} - \mathbf{rq})\mathbf{p}$$

$$= 4\mathbf{p} \times (\mathbf{q} \times \mathbf{r}).$$

As an immediate consequence, we find:

1.17   Sum Identity for the Double Vector Product

Let $\mathbf{p}$, $\mathbf{q}$ and $\mathbf{r}$ be three pure quaternions. Then it holds that

$$\mathbf{p} \times (\mathbf{q} \times \mathbf{r}) + \mathbf{q} \times (\mathbf{r} \times \mathbf{p}) + \mathbf{r} \times (\mathbf{p} \times \mathbf{q}) = 0_{\mathbb{H}}.$$

Indeed, if we apply the development formula successively to each of the three products, we get

$$\mathbf{p} \times (\mathbf{q} \times \mathbf{r}) + \mathbf{q} \times (\mathbf{r} \times \mathbf{p}) + \mathbf{r} \times (\mathbf{p} \times \mathbf{q}) = \mathbf{q}(\mathbf{p} \cdot \mathbf{r}) - \mathbf{r}(\mathbf{p} \cdot \mathbf{q}) + \mathbf{r}(\mathbf{q} \cdot \mathbf{p})$$

$$- \mathbf{p}(\mathbf{q} \cdot \mathbf{r}) + \mathbf{p}(\mathbf{r} \cdot \mathbf{q}) - \mathbf{q}(\mathbf{r} \cdot \mathbf{p})$$

$$= 0_{\mathbb{H}}.$$

These results make it plain that it is also useful to formulate a vector form of the well-known Lagrange identity.

1.18   Lagrange Identity

Let $\mathbf{p}$, $\mathbf{q}$, $\mathbf{r}$ and $\mathbf{s}$ be four pure quaternions. For the scalar product of two vector products we have

$$(\mathbf{p} \times \mathbf{q}) \cdot (\mathbf{r} \times \mathbf{s}) = \det \begin{pmatrix} \mathbf{p} \cdot \mathbf{r} & \mathbf{p} \cdot \mathbf{s} \\ \mathbf{q} \cdot \mathbf{r} & \mathbf{q} \cdot \mathbf{s} \end{pmatrix}.$$

To see this, we note that

$$(\mathbf{p} \times \mathbf{q}) \cdot (\mathbf{r} \times \mathbf{s}) = [(\mathbf{p} \times \mathbf{q}) \times \mathbf{r}] \cdot \mathbf{s} = [\mathbf{q}(\mathbf{p} \cdot \mathbf{r}) - \mathbf{p}(\mathbf{q} \cdot \mathbf{r})] \cdot \mathbf{s}$$

$$= (\mathbf{q} \cdot \mathbf{s})(\mathbf{p} \cdot \mathbf{r}) - (\mathbf{p} \cdot \mathbf{s})(\mathbf{q} \cdot \mathbf{r}).$$

Example. Prove that

$$(\mathbf{p} \times \mathbf{q}) \times (\mathbf{r} \times \mathbf{s}) = \mathbf{q}(\mathbf{p} \cdot (\mathbf{r} \times \mathbf{s})) - \mathbf{p}(\mathbf{q} \cdot (\mathbf{r} \times \mathbf{s})) = \mathbf{q}(\mathbf{p}, \mathbf{r}, \mathbf{s}) - \mathbf{p}(\mathbf{q}, \mathbf{r}, \mathbf{s})$$

holds for any pure quaternions $\mathbf{p}$, $\mathbf{q}$, $\mathbf{r}$ and $\mathbf{s}$.

Solution. Indeed, using (1.3), we have

$$(\mathbf{p} \times \mathbf{q}) \times (\mathbf{r} \times \mathbf{s}) = -(\mathbf{r} \times \mathbf{s}) \times (\mathbf{p} \times \mathbf{q})$$

$$= -\left[((\mathbf{r} \times \mathbf{s}) \cdot \mathbf{q})\,\mathbf{p} - ((\mathbf{r} \times \mathbf{s}) \cdot \mathbf{p})\,\mathbf{q}\right]$$

$$= \mathbf{q}\,(\mathbf{p} \cdot (\mathbf{r} \times \mathbf{s})) - \mathbf{p}\,(\mathbf{q} \cdot (\mathbf{r} \times \mathbf{s}))$$

$$= \mathbf{q}(\mathbf{p}, \mathbf{r}, \mathbf{s}) - \mathbf{p}(\mathbf{q}, \mathbf{r}, \mathbf{s}).$$

Example. Let $\mathbf{p}$, $\mathbf{q}$ and $\mathbf{r}$ be pure quaternions. Prove the double factor rule

$$(\mathbf{p} \times \mathbf{q}) \times (\mathbf{q} \times \mathbf{r}) = \mathbf{q}(\mathbf{p}, \mathbf{q}, \mathbf{r}).$$

Solution. Here we use again (1.3). A direct computation shows that

$$(\mathbf{p} \times \mathbf{q}) \times (\mathbf{q} \times \mathbf{r}) = ((\mathbf{p} \times \mathbf{q}) \cdot \mathbf{r})\,\mathbf{q} - ((\mathbf{p} \times \mathbf{q}) \cdot \mathbf{q})\,\mathbf{r}$$
$$= \mathbf{q}\,((\mathbf{p} \times \mathbf{q}) \cdot \mathbf{r})$$
$$= \mathbf{q}\,(\mathbf{r} \cdot (\mathbf{p} \times \mathbf{q}))$$
$$= \mathbf{q}(\mathbf{r}, \mathbf{p}, \mathbf{q}).$$

1.19 Quaternion Outer and Even Products

The quaternion outer product of p and q is defined by

$$(p, q) := p_0 \mathbf{q} - q_0 \mathbf{p} - \mathbf{p} \times \mathbf{q}.$$

The even product of quaternions is not widely used either, but it is mentioned here due to its similarity with the classical odd product; it is given by

$$[p, q] := p_0 q_0 - \mathbf{p}\mathbf{q} + p_0 \mathbf{q} + q_0 \mathbf{p}.$$

This is the purely symmetric product. Therefore, it is commutative.

Example. Find (p, q) and $[p, q]$, where $p = 1 - i - j - k$ and $q = j + k$.

Solution. Direct calculations show that

$$(p, q) = j + k + (i + j + k) \times (j + k) = j + k - j + k = 2k,$$

$$[p, q] = -(-i - j - k)(j + k) + j + k = -2 + 2k.$$

Exercise 1.36. *Let $p = 1 - i - k$ and $q = j + k$. Find (p, q) and $[p, q]$.*

Solution. $(p, q) = -i - j + k,\ [p, q] = 1 + i + j - k.$

1.20 Equivalent Quaternions

The real quaternions p_1 and p_2 are said to be equivalent, denoted by $p_1 \sim p_2$, if there exists an $h \in \mathbb{H} \setminus \{0_{\mathbb{H}}\}$ such that $h p_1 = p_2 h$.

Example. Check whether $i \sim j$.

Solution. We want to find a quaternion $h = a_1 + b_1 i + c_1 j + d_1 k$ such that

$$(a_1 + b_1 i + c_1 j + d_1 k)i = j(a_1 + b_1 i + c_1 j + d_1 k),$$

which is equivalent to

$$-b_1 + a_1 i + d_1 j - c_1 k = -c_1 + d_1 i + a_1 j - b_1 k.$$

It follows that $a_1 = d_1$ and $b_1 = c_1$. If $a_1, b_1 \in \mathbb{R} \setminus \{0\}$, we have that $i \sim j$.

Exercise 1.37. *Check whether* (a) $i \sim k$,   (b) $j \sim k$.

Solution. (a) $h = a_1 + b_1 i - a_1 j + b_1 k$, with $a_1, b_1 \in \mathbb{R} \setminus \{0\}$;   (b) $h = a_1 + a_1 i + c_1 j + c_1 k$, with $a_1, c_1 \in \mathbb{R} \setminus \{0\}$.

Exercise 1.38. *Check whether* $1 + i - j + k \sim i + j$.

Solution. These quaternions are not equivalent.

Exercise 1.39. *Let* $p_1 = a_1 + b_1 i + c_1 j + d_1 k$ *and* $p_2 = a_2 + b_2 i + c_2 j + d_2 k$ *with* $a_i, b_i, c_i, d_i \in \mathbb{R}$ $(i = 1, 2)$. *Determine* a_i, b_i, c_i *and* d_i $(i = 1, 2)$ *so that* $p_1 \sim p_2$.

Solution. $p_1 \sim p_2$ if

$$\det \begin{pmatrix} a_1 - a_2 & b_2 - b_1 & c_2 - c_1 & d_2 - d_1 \\ b_1 - b_2 & a_1 - a_2 & d_2 - d_1 & -c_1 - c_2 \\ c_1 - c_2 & -d_1 - d_2 & a_1 - a_2 & b_1 + b_2 \\ d_1 - d_2 & c_1 + c_2 & -b_1 - b_2 & a_1 - a_2 \end{pmatrix} = 0.$$

Exercise 1.40. *Let* $p_1 = a_1 + b_1 i + c_1 j + d_1 k$, *where* $a_1, b_1, c_1, d_1 \in \mathbb{R} \setminus \{0\}$. *Prove that* $a_1 + \sqrt{b_1^2 + c_1^2 + d_1^2}\, i \sim p_1$.

Hint. Use the previous exercise.

Example. Let p_1 and p_2 be real quaternions. Prove that $p_1 \sim p_2$ if and only if $\mathrm{Sc}(p_1) = \mathrm{Sc}(p_2)$ and $|p_1| = |p_2|$.

Solution. We first prove the necessity part. Let $p_1 \sim p_2$. Then there exists a nonnull quaternion h such that $hp_1 = p_2 h$. It follows that $|hp_1| = |p_2 h|$, that is, $|h||p_1| = |p_2||h| = |h||p_2|$. Since $h \neq 0_{\mathbb{H}}$ then $|h| \neq 0$. Hence $|p_1| = |p_2|$. Also, $p_1 = h^{-1} p_2 h$. One can verify directly that

$$\mathrm{Sc}(p_1) = \mathrm{Sc}((h^{-1} p_2)h) = \mathrm{Sc}(h(h^{-1} p_2)) = \mathrm{Sc}(p_2).$$

For the sufficiency part, we assume that there are quaternions p_1 and p_2 such that $|p_1| = |p_2|$ and $\mathrm{Sc}(p_1) = \mathrm{Sc}(p_2)$. Let $p_1 = a_1 + b_1 i + c_1 j + d_1 k$ and $p_2 =$

$a_1 + b_2 i + c_2 j + d_2 k$, with $a_i, b_i, c_i, d_i \in \mathbb{R} \setminus \{0\}$ $(i = 1, 2)$. Then $b_1^2 + c_1^2 + d_1^2 = b_2^2 + c_2^2 + d_2^2$. From the previous exercise we have

$$a_1 + \sqrt{b_1^2 + c_1^2 + d_1^2}\, i \sim p_1,$$

$$a_1 + \sqrt{b_2^2 + c_2^2 + d_2^2}\, i = a_1 + \sqrt{b_1^2 + c_1^2 + d_1^2}\, i \sim p_2.$$

Therefore, $p_1 \sim p_2$.

Example. Let p_1 and p_2 be real quaternions, and n a natural number. Prove that $\left(p_1^{-1} p_2 p_1\right)^n \sim p_2^n$.

Solution. We will use mathematical induction to show that $(p_1^{-1} p_2 p_1)^n = p_1^{-1} p_2^n p_1$ holds for every natural n. For $n = 1$, the relation is obvious. When $n = 2$ we have

$$\left(p_1^{-1} p_2 p_1\right)^2 = p_1^{-1} p_2 p_1 p_1^{-1} p_2 p_1 = p_1^{-1} p_2^2 p_1.$$

Now we suppose that $\left(p_1^{-1} p_2 p_1\right)^n = p_1^{-1} p_2^n p_1$ holds, and prove that

$$\left(p_1^{-1} p_2 p_1\right)^{n+1} = p_1^{-1} p_2^{n+1} p_1.$$

Straightforward computations show that

$$\left(p_1^{-1} p_2 p_1\right)^{n+1} = \left(p_1^{-1} p_2 p_1\right)^n p_1^{-1} p_2 p_1 = p_1^{-1} p_2^n p_1 p_1^{-1} p_2 p_1 = p_1^{-1} p_2^{n+1} p_1.$$

Consequently, $\left(p_1^{-1} p_2 p_1\right)^n = p_1^{-1} p_2^n p_1$ for every natural n.

The above definitions centered most use of quaternions around the vector part, while neglecting the scalar part. Some of the neat mathematical characteristics of quaternions are indeed their algebraic properties (like division, which is missing for modern vectors). In fact, the properties of the multiplication and division of quaternions and the law of the moduli enable one to treat quaternions as numbers, in the same way as complex numbers. This makes it possible to do interesting research, e.g. to derive closed form solutions for algebraic systems involving unknown rotational parameters. This contributed to the development of the modern theory of vectors.

We proceed by recalling the notion of polar coordinate system from calculus, firstly invented in 1637 by René Descartes (1596–1650), and taken up several decades after by Sir Isaac Newton (1642–1726) and Daniel Bernoulli (1700–1782), among others. The polar coordinate system consists of a two-dimensional coordinate system in which each point P on a plane whose rectangular coordinates are (x, y) is determined by the distance from a fixed point and the angle made with a fixed direction. The fixed point O (analogous to the origin of a Cartesian system) is called the pole, and the horizontal half-line emanating from the pole with the fixed direction is the polar axis. If r is a directed distance from the pole to P, known as radial coordinate or radius, and the polar angle θ (also known as angular coordinate or azimuth) is the angle of rotation measured from the polar axis to the line OP, then the point P can be described by the ordered pair (r, θ), called the polar coordinates of P.

1.21   Polar Form of a Quaternion

Let p be a real quaternion such that $\mathbf{p} \neq (0, 0, 0)$. The quaternion p can also be associated with an angle θ via

$$\cos\theta = \frac{p_0}{|p|} \quad \text{and} \quad \sin\theta = \frac{|\mathbf{p}|}{|p|}.$$

This superimposed argument is in some sense correct since, obviously,

$$-1 \le \cos\theta \le 1, \quad -1 \le \sin\theta \le 1, \quad \text{and} \quad \cos^2\theta + \sin^2\theta = 1.$$

Every nonzero quaternion p can be written in polar form[5]

$$p = |p|\left(\cos\theta + \frac{\mathbf{p}}{|\mathbf{p}|}\sin\theta\right). \tag{1.4}$$

We say that (1.4) is the polar representation of the quaternion p. In a polar coordinate system the angle θ is viewed as the angle between the vector p in $\mathbb{R}^4$ and the real axis (the subspace of real numbers), and $\frac{|\mathbf{p}|}{|p|}\sin\theta$ as the projection of p onto the subspace $\mathbb{R}^3$ of pure quaternions. Furthermore, the angle θ is always measured in radians, and is positive when measured counterclockwise and negative when measured clockwise. As usual, radians can conveniently express angle measures in terms of π. For example, an angle of $\frac{\pi}{2}$ radians is equal to an angle of $90°$ and an angle of π radians is equal to an angle of $180°$.

For any $p \in \mathbb{H}$ with $\mathbf{p} \neq (0, 0, 0)$, we clearly have the equality

$$p = |p|\left(\frac{p_0}{|p|} + \frac{\mathbf{p}}{|p|}\right) = |p|\left(\frac{p_0}{|p|} + \frac{\mathbf{p}}{|\mathbf{p}|}\frac{|\mathbf{p}|}{|p|}\right).$$

This can be easily seen by observing that $p_0^2 + |\mathbf{p}|^2 = |p|^2$, since there is a unique angle θ such that

---

[5]It is also known as trigonometric representation of the quaternion p.

$$|p|\left(\cos\theta + \frac{\mathbf{p}}{|\mathbf{p}|}\sin\theta\right).$$

Example. Find the trigonometric representation of $p = 1 - i + 2j + k$.

Solution. It follows that $\theta = \arccos(1/\sqrt{7}) = \arcsin(\sqrt{6/7})$. In this way we obtain the representation

$$p = \sqrt{7}\left(\frac{1}{\sqrt{7}} + \frac{-i+2j+k}{\sqrt{6}}\sqrt{\frac{6}{7}}\right).$$

Exercise 1.41. *Find the trigonometric representation of the quaternions:* (a) $p = 1 + i + j + k$; (b) $p = j + k$; (c) $p = i + k$; (d) $p = 1 + i + k$; (e) $p = 1 - i - j$.

Solution. (a) $2\left(\frac{1}{2} + \frac{i+j+k}{\sqrt{3}}\frac{\sqrt{3}}{2}\right)$; (b) $\sqrt{2}\left(0 + \frac{j+k}{\sqrt{2}}1\right)$; (c) $\sqrt{2}\left(0 + \frac{i+k}{\sqrt{2}}1\right)$; (d) $\sqrt{3}\left(\frac{1}{\sqrt{3}} + \frac{i+k}{\sqrt{2}}\sqrt{\frac{2}{3}}\right)$; (e) $\sqrt{3}\left(\frac{1}{\sqrt{3}} - \frac{i+j}{\sqrt{2}}\sqrt{\frac{2}{3}}\right)$.

Exercise 1.42. *Let $p = i + j + k$ and $q = j - k$. Using the trigonometric representation compute:* (a) p^2; (b) pq; (c) $p - q$.

Solution. (a) -3; (b) $-2i + j + k$; (c) $i + 2k$.

Exercise 1.43. *Using the polar form of a quaternion p, evaluate $p^n + \overline{p}^n$ and $p^n - \overline{p}^n$ for $n \in \mathbb{N}$.*

Solution. $p^n + \overline{p}^n = 2|p|^n\cos(n\theta)$ and $p^n - \overline{p}^n = 2|p|^n\frac{\mathbf{p}}{|\mathbf{p}|}\sin(n\theta)$ for some angle θ.

1.22 Quaternion Sign and Quaternion Argument

The quaternion sign produces the unit quaternion, $\operatorname{sgn}(p) = \frac{p}{|p|}$. The angle θ is called the quaternion argument or amplitude of the quaternion p and is denoted by $\theta := \arg(p)$. An argument can be assigned to any nonzero real quaternion p. However, for $p = 0_{\mathbb{H}}$, $\arg(p)$ cannot be defined in any way that is meaningful. Also, an argument of a real quaternion p is not unique, since $\cos\theta$ and $\sin\theta$ are 2π-periodic functions. That means if θ_0 is an argument of p, then necessarily the angles $\theta_0 \pm 2\pi n$ for any integer n are also arguments of p, and we can sometimes think of $\arg(p)$ as $\{\theta \pm 2\pi n : n \in \mathbb{N}\}$. In practice, when $p \neq 0_{\mathbb{H}}$ we use $\tan\theta = \frac{|\mathbf{p}|}{p_0}$ to find θ, where the quadrant in which p_0 and $|\mathbf{p}|$ lie must always be specified or be clearly understood.

1.23   Quaternion Argument of a Product

The polar form of a quaternion is especially convenient when multiplying two pure quaternions with proportional vector parts. If we consider a unit pure quaternion p (i.e. such that $|\mathbf{p}| = 1$), then it is undoubtedly clear that $p^2 = -1_{\mathbb{H}}$, as has already been discussed. Therefore, the product of two unit pure quaternions with proportional vector parts $p = |p|(\cos\theta + p\sin\theta)$ and $q = |q|(\cos\phi + p\sin\phi)$ yields the following de Moivre's type formula for pure quaternions:

$$pq = |p||q|\left(\cos(\theta + \phi) + p\sin(\theta + \phi)\right).$$

From this it is evident that the argument of pq is given by $\arg(p + q) = \arg(p) + \arg(q)$.

1.24   Principal Argument

Although the symbol $\arg(p)$ actually represents a set of values, the argument θ of a quaternion that lies in the interval $0 \leq \theta \leq \pi$ is called (by analogy with the complex case) the principal value of $\arg(p)$ or the principal argument of p. The principal argument of p is unique and is denoted by the symbol $\mathrm{Arg}(p)$, that is, $\mathrm{Arg}(p) \in [0, \pi]$. In general, $\arg(p)$ and $\mathrm{Arg}(p)$ are related by $\arg(p) = \mathrm{Arg}(p) + 2\pi n, n = 0, \pm 1, \pm 2, \ldots$. For example, $\arg(i + j + k) = \frac{\pi}{2} + 2\pi n, n \in \mathbb{Z}$. For the choices $n = 0$ and $n = 1$, it follows that $\arg(i + j + k) = \mathrm{Arg}(i + j + k) = \frac{\pi}{2}$ and $\arg(i + j + k) = \frac{5\pi}{2}$, respectively.

1.25   de Moivre's Formula

Let p be a quaternion such that $\mathbf{p} \neq (0, 0, 0)$, and θ is real. Then, for $n = 0, \pm 1, \pm 2, \ldots$ we have

$$\left(\cos\theta + \mathrm{sgn}(\mathbf{p})\sin\theta\right)^n = \cos(n\theta) + \mathrm{sgn}(\mathbf{p})\sin(n\theta). \tag{1.5}$$

Contrary to the complex case, here we must replace the imaginary unit i by an element of the two-dimensional unit sphere. But we nevertheless have $\mathrm{sgn}^2(\mathbf{p}) = -1_{\mathbb{H}}$. For positive n the proof then proceeds by mathematical induction

$$\left(\cos\theta + \mathrm{sgn}(\mathbf{p})\sin\theta\right)^{n+1}$$

$$= \left(\cos(n\theta) + \mathrm{sgn}(\mathbf{p})\sin(n\theta)\right)\left(\cos\theta + \mathrm{sgn}(\mathbf{p})\sin\theta\right)$$

$$= \cos\left((n + 1)\theta\right) + \mathrm{sgn}(\mathbf{p})\sin\left((n + 1)\theta\right).$$

The case $n = 0$ gives $1_\mathbb{H}$ on both sides of (1.5), and for negative n we have the relation

$$\left(\cos\theta + \text{sgn}(\mathbf{p})\sin\theta\right)^{-n} = \left(\cos\theta + \text{sgn}(\mathbf{p})\sin\theta\right)^{n}$$

$$= \cos(n\theta) - \text{sgn}(\mathbf{p})\sin(n\theta)$$

$$= \cos(-n\theta) + \text{sgn}(\mathbf{p})\sin(-n\theta).$$

Exercise 1.44. *Find all quaternions p such that:* (a) $p^3 = 0_\mathbb{H}$;   (b) $p^4 = 1_\mathbb{H}$.

Solution. (a) $p = 0_\mathbb{H}$;   (b) $p = 1_\mathbb{H}$.

Exercise 1.45. *Find all quaternions p such that:* (a) $p^n + \overline{p}^n = 1_\mathbb{H}$ *for $n \in \mathbb{N}$;* (b) $p^n - \overline{p}^n = 1_\mathbb{H}$ *for $n \in \mathbb{N}$.*

Solution. (a) all quaternions p;   (b) no solutions.

1.26   Failure for Noninteger Powers

de Moivre's formula does not hold in general for noninteger powers. As a brief preview of the material in Chap. 5, we shall notice that noninteger powers of a real quaternion may have many different values. To see this, let us assume that

$$p = |p|\left(\cos\theta + \frac{\mathbf{p}}{|\mathbf{p}|}\sin\theta\right) \in \mathbb{H}$$

so that $\mathbf{p} \neq (0,0,0)$. Then

$$\sqrt[k]{p} = \sqrt[k]{|p|}\left(\cos\frac{(\theta + 2\pi n)}{k} + \text{sgn}(\mathbf{p})\sin\frac{(\theta + 2\pi n)}{k}\right),$$

for $n = 0, \ldots, k-1$. As n takes on the successive integer values $n = 0, 1, 2, \ldots, k-1$ we obtain k distinct kth roots of p; these roots have the same modulus $\sqrt[k]{|p|}$, but different arguments. Consider, for instance, the quaternion $p = 2i - 2j + k$. It follows that $|p| = 3$, $\text{sgn}(p) = \frac{2i - 2j + k}{3}$, and $\theta = \frac{\pi}{2}$. This yields

$$\sqrt{p} = \sqrt{3}\left(\cos\frac{(\pi/2 + 2\pi n)}{2} + \frac{(2i - 2j + k)}{3}\sin\frac{(\pi/2 + 2\pi n)}{2}\right),$$

for $n = 0, 1$.

Exercise 1.46. *Compute the indicated powers of the quaternion $p = -2 + i - \frac{1}{3}j + 5k$:* (a) $p^{\frac{1}{3}}$;   (b) $p^{\frac{1}{7}}$;   (c) p^{11};   (d) p^{23};   (e) p^n *for $n \in \mathbb{N}$.*

Solution. $p = \frac{\sqrt{271}}{3}\left[\cos(\arccos(-\frac{6}{\sqrt{271}})) + \frac{3\,(i-\frac{1}{3}j+5k)}{\sqrt{235}}\sin(\arccos(-\frac{6}{\sqrt{271}}))\right].$

In the preceding part of text we have seen that quaternions are indeed related to both real and complex numbers. Just as complex numbers can be represented as matrices, so can quaternions. One of the effective approaches in studying matrices of quaternions is to convert a given real quaternion into a 2×2 complex matrix, and the other is to use a 4×4 real matrix. In the terminology and methodology of abstract algebra, these are injective homomorphisms from $\mathbb{H}$ to the matrix rings $\mathcal{M}_2(\mathbb{C})$ and $\mathcal{M}_4(\mathbb{R})$, respectively. The first group is important for describing spin in quantum mechanics, which is intimately related with the widely known Pauli matrices.[6] Although they visualize quaternions well (see the matrix representation below), the latter representation is not particularly useful because 4×4 matrices require often much more operations than 2×2 matrices.

1.27   First Matrix Representation of Quaternions

A real quaternion p can be represented using 2×2 complex matrices as

$$p := a + bi + cj + dk \quad \leftrightarrow \quad \mathcal{M}_p(\mathbb{C}) = a\,\sigma_0 + b\,\sigma_1 + c\,\sigma_2 + d\,\sigma_3$$

$$:= \begin{pmatrix} a+bi & c+di \\ -c+di & a-bi \end{pmatrix} \tag{1.6}$$

where σ_0 is the 2×2 identity matrix and the σ_l with $l = 1, 2, 3$, are the Hermitian Pauli spin matrices,

$$\sigma_0 = \begin{pmatrix} 1 & 0 \\ 0 & 1 \end{pmatrix}, \quad \sigma_1 = \begin{pmatrix} i & 0 \\ 0 & -i \end{pmatrix}, \quad \sigma_2 = \begin{pmatrix} 0 & 1 \\ -1 & 0 \end{pmatrix}, \quad \sigma_3 = \begin{pmatrix} 0 & i \\ i & 0 \end{pmatrix}.$$

A straightforward computation shows that

$$\sigma_1^2 = \sigma_2^2 = \sigma_3^2 = -\sigma_0,$$

$$\sigma_1\sigma_2 = \sigma_3 = -\sigma_2\sigma_1, \quad \sigma_2\sigma_3 = \sigma_1 = -\sigma_3\sigma_2, \quad \sigma_3\sigma_1 = \sigma_2 = -\sigma_1\sigma_3.$$

The representation (1.6) has the following properties:
  (i) Complex numbers ($c = d = 0$) correspond to diagonal matrices;
  (ii) The absolute value of p is the square root of the determinant of the corresponding matrix $\mathcal{M}_p(\mathbb{C})$;

---

[6]These matrices are used in physics to describe the angular momentum of elementary particles such as electrons.

(iii) The conjugate quaternion $\bar{p}$ corresponds to the conjugate transpose of the corresponding matrix $\mathcal{M}_p(\mathbb{C})$;

(iv) If p is a unit quaternion, then $\det(\mathcal{M}_p(\mathbb{C})) = 1$ and $\mathcal{M}_p(\mathbb{C})$ is a unitary matrix (i.e. $\mathcal{M}_p(\mathbb{C})\overline{\mathcal{M}_p(\mathbb{C})}^T = \sigma_0$).

Example. Let $p = 2 - j + k$ and $q = -1 + 2i + j - k$. Find pq using the first matrix representation.

Solution. The first matrix representations of $p = 2 - j + k$ and $q = -1 + 2i + j - k$ are, respectively,

$$\mathcal{M}_p(\mathbb{C}) := \begin{pmatrix} 2 & -1+i \\ 1+i & 2 \end{pmatrix} \quad \text{and} \quad \mathcal{M}_q(\mathbb{C}) := \begin{pmatrix} -1+2i & 1-i \\ -1-i & -1-2i \end{pmatrix}.$$

The first matrix representation of the product of p and q is then given by

$$\mathcal{M}_p(\mathbb{C})\,\mathcal{M}_q(\mathbb{C}) = \begin{pmatrix} 2 & -1+i \\ 1+i & 2 \end{pmatrix}\begin{pmatrix} -1+2i & 1-i \\ -1-i & -1-2i \end{pmatrix}$$

$$= \begin{pmatrix} 4i & 5-i \\ -5-i & -4i \end{pmatrix}.$$

That is, $pq = 4i + 5j - k$.

Exercise 1.47. *Let $p = 24i + 4j + 4k$ and $q = ijk$. Compute $p^2(1 - q^2)$ using the first matrix representation.*

Solution. $0_{\mathbb{H}}$.

Exercise 1.48. *Represent the following quaternions by the first matrix representation:* (a) $p = -i + j + k$; (b) $p = 5i + k$; (c) $p = 1 + i - 3k$; (d) $p = i(-j + k)(11 - 3k)$.

Solution. (a) $\begin{pmatrix} -i & 1+i \\ -1+i & i \end{pmatrix}$; (b) $\begin{pmatrix} 5i & i \\ i & -5i \end{pmatrix}$; (c) $\begin{pmatrix} 1+i & -3i \\ -3i & 1-i \end{pmatrix}$;

(d) $\begin{pmatrix} -3+3i & -11-11i \\ 11-11i & -3-3i \end{pmatrix}$.

1.28 Second Matrix Representation of Quaternions

A real quaternion p can be represented using 4×4 real matrices as

$$p := a + bi + cj + dk \quad \leftrightarrow \quad \mathcal{M}_p^{(1)}(\mathbb{R}) := \begin{pmatrix} a & b & c & d \\ -b & a & -d & c \\ -c & d & a & -b \\ -d & -c & b & a \end{pmatrix}. \tag{1.7}$$

The representation (1.7) has the following properties:

(i) $\mathcal{M}_{1_{\mathbb{H}}}^{(1)}(\mathbb{R}) = I_4$, where I_4 denotes the identity matrix of order 4;

(ii) Complex numbers ($c = d = 0$) are block diagonal matrices with two 2×2 blocks;

(iii) The fourth power of the absolute value of p is the determinant of the corresponding matrix $\mathcal{M}_p^{(1)}(\mathbb{R})$;

(iv) The conjugate quaternion $\overline{p}$ corresponds to the transpose of the corresponding matrix $\mathcal{M}_p^{(1)}(\mathbb{R})$;

(v) If p is a unit quaternion, then $\det(\mathcal{M}_p^{(1)}(\mathbb{R})) = 1$ and $\mathcal{M}_p^{(1)}(\mathbb{R})$ is an orthogonal matrix (i.e. $\mathcal{M}_p^{(1)}(\mathbb{R})\mathcal{M}_p^{(1)}(\mathbb{R})^T = I_4$).

The verification of these statements is not difficult and is left to the reader. In particular, the matrix $\mathcal{M}_p^{(1)}(\mathbb{R})$ enjoys the properties:

(vi) $\mathcal{M}_p^{(1)}(\mathbb{R})^T \mathcal{M}_p^{(1)}(\mathbb{R}) = \mathcal{M}_p^{(1)}(\mathbb{R})\mathcal{M}_p^{(1)}(\mathbb{R})^T = |p|^2 I_4$;

(vii) $\mathcal{M}_{pq}^{(1)}(\mathbb{R}) = \mathcal{M}_p^{(1)}(\mathbb{R})\,\mathcal{M}_q^{(1)}(\mathbb{R})$.

Exercise 1.49. *Show that the matrix associated with a pure quaternion p is antisymmetric, and it satisfies the following property:* $\mathcal{M}_p^{(1)}(\mathbb{R})^T = -\mathcal{M}_p^{(1)}(\mathbb{R})$.

Exercise 1.50. *Let* (a) $p = 1 + i + j$;   (b) $p = 1 - i - k$;   (c) $p = 2 + j + k$;   (d) $p = 1 - i - j - k$;   (e) $p = j + k$. *Find the quaternions that correspond to the following matrices:*

$$i A^2 + jA + (1 - i - j)I_3 \quad \text{and} \quad i B^2 + jB + (1 - i - j)I_4,$$

where A and B are, respectively, the first and second matrix representations of p.

Solution. (a) $-2 - 2i + k$;   (b) $3 - 3i + 2j + k$;   (c) $-3j + 4k$;   (d) $5 - 4i + 2j - k$;   (e) $-2i - j$.

Exercise 1.51. *Let $p \in \mathbb{H}$. Find the first and second matrix representations of p_+ and p_-.*

Solution.

$$\mathcal{M}_{p+}(\mathbb{C}) = \begin{pmatrix} \frac{a+d}{2} + \frac{b-c}{2}i & \frac{c-b}{2} + \frac{a+d}{2}i \\ \frac{b-c}{2} + \frac{a+d}{2}i & \frac{a+d}{2} - \frac{b-c}{2}i \end{pmatrix},$$

$$\mathcal{M}_{p+}^{(1)}(\mathbb{R}) = \begin{pmatrix} \frac{a+d}{2} & \frac{c-b}{2} & \frac{a+d}{2} & \frac{b-c}{2} \\ \frac{b-c}{2} & \frac{a+d}{2} & \frac{b-c}{2} & -\frac{a+d}{2} \\ -\frac{a+d}{2} & \frac{c-b}{2} & \frac{a+d}{2} & \frac{c-b}{2} \\ \frac{c-b}{2} & \frac{a+d}{2} & \frac{b-c}{2} & \frac{a+d}{2} \end{pmatrix},$$

$$\mathcal{M}_{p-}(\mathbb{C}) = \begin{pmatrix} \frac{a-d}{2} + \frac{b+c}{2}i & \frac{b+c}{2} + \frac{d-a}{2}i \\ -\frac{b+c}{2} + \frac{d-a}{2}i & \frac{a-d}{2} - \frac{b+c}{2}i \end{pmatrix},$$

$$\mathcal{M}_{p-}^{(1)}(\mathbb{R}) = \begin{pmatrix} \frac{a-d}{2} & -\frac{c+b}{2} & \frac{d-a}{2} & -\frac{b+c}{2} \\ \frac{b+c}{2} & \frac{a-d}{2} & -\frac{b+c}{2} & -\frac{d-a}{2} \\ -\frac{d-a}{2} & \frac{c+b}{2} & \frac{a-d}{2} & -\frac{c+b}{2} \\ \frac{c+b}{2} & \frac{d-a}{2} & \frac{b+c}{2} & \frac{a-d}{2} \end{pmatrix}.$$

Other matrix representations are possible. In a completely analogous way, we may also consider another quaternion matrix representation:

$$\mathcal{M}_{p}^{(2)}(\mathbb{R}) := \begin{pmatrix} a & -b & d & -c \\ b & a & -c & -d \\ -d & c & a & -b \\ c & d & b & a \end{pmatrix}.$$

This matrix satisfies the following properties:

(i) $\mathcal{M}_{p}^{(2)}(\mathbb{R})^{T}\mathcal{M}_{p}^{(2)}(\mathbb{R}) = \mathcal{M}_{p}^{(2)}(\mathbb{R})\mathcal{M}_{p}^{(2)}(\mathbb{R})^{T} = |p|^{2}\,I_{4}$;

(ii) $\mathcal{M}_{pq}^{(2)}(\mathbb{R}) = \mathcal{M}_{p}^{(2)}(\mathbb{R})\,\mathcal{M}_{q}^{(2)}(\mathbb{R})$;

(iii) $\det(\mathcal{M}_{p}^{(2)}(\mathbb{R})) = |p|^{4}$.

Exercise 1.52. *Prove the validity of the previous statements.*

Exercise 1.53. *Prove that* $\mathcal{M}_{i}^{(2)}(\mathbb{R})\mathcal{M}_{j}^{(2)}(\mathbb{R})\mathcal{M}_{k}^{(2)}(\mathbb{R}) = -I_{4}$.

For the reader's convenience, from now on we only deal with the above right second matrix representation, $\mathcal{M}_{p}^{(2)}(\mathbb{R})$, which for simplicity we call second matrix representation. The other cases may be treated analogously.

1.29 Advanced Practical Exercises

1. Let $p = 3 + i$ and $q = 5i + j - 2k$. Compute (a) $p + q$;   (b) $p - q$;
   (c) $2p - 7q$;   (d) $3p + 4q$.

2. Compute and compare the results: (a) $i(jk)$, $(ij)k$; (b) $j(ik)$, $(ji)k$; (c) $i(kj)$, $(ik)j$; (d) $j(ki)$, $(jk)i$; (e) $k(ij)$, $(ki)j$; (f) $k(ji)$, $(kj)i$.

3. Let $p = 1 + i - j - k$, $q = 3 + 2i + j + 4k$, and $r = 2 + i + j + k$. Compute (a) pq; (b) pr; (c) qr; (d) pqr; (e) p^2; (f) q^3; (g) pqr^2; (h) $2p - pqr^2$; (i) $p^2 + q$; (j) $p + q^3 + pqr^2$.

4. Find $p = a + bi + cj + dk$ $(a, b, c, d \in \mathbb{R})$ such that: (a) $p(1 + i - j) = 1_{\mathbb{H}}$, $(1 + i - j)p = 1_{\mathbb{H}}$; (b) $pi = 1_{\mathbb{H}}$, $ip = 1_{\mathbb{H}}$; (c) $pj = 1_{\mathbb{H}}$, $jp = 1_{\mathbb{H}}$; (d) $pk = 1_{\mathbb{H}}$, $kp = 1_{\mathbb{H}}$.

5. Let a, b, c, d be given real numbers for which $(a, b, c, d) \neq (0, 0, 0, 0)$. Find $p = a_1 + b_1 i + c_1 j + d_1 k$ $(a_1, b_1, c_1, d_1 \in \mathbb{R})$, such that: (a) $(a + bi + cj + dk)p = 1_{\mathbb{H}}$; (b) $p(a + bi + cj + dk) = 1_{\mathbb{H}}$.

6. Find a solution $p = a + bi + cj + dk$ $(a, b, c, d \in \mathbb{R})$, to the following equations: (a) $2p = 1 - i + j + 3k$; (b) $p^2 - p = 1_{\mathbb{H}}$; (c) $p^2 - p = i$; (d) $p^2 - p = j$; (e) $p^2 - p = k$; (f) $p(1 + i) = i + j + k$; (g) $(1 + i)p = i + j + k$; (h) $(1 - 2i + j + k)p = 1 + 3i + j + k$; (i) $p^2 = -1_{\mathbb{H}}$; (j) $p^2 + ip = 1_{\mathbb{H}}$; (k) $p^2 + ip = j$; (l) $p^2 + ip = k$.

7. Let $p = 2 - i + j$ and $q = 1 + i + j + k$. Compute: (a) $\overline{p}$; (b) $\overline{q}$; (c) $\overline{pq}$; (d) $\overline{p}\,\overline{q}$; (e) $\overline{q}\,\overline{p}$; (f) $\overline{qp}$.

8. Let $p = a_1 + b_1 i + c_1 j + d_1 k$ and $q = a_2 + b_2 i + c_2 j + d_2 k$. Compute: (a) $\overline{pq}$; (b) $\overline{p}\,\overline{q}$; (c) $\overline{q}\,\overline{p}$; (d) $\overline{qp}$.

9. Check that $\overline{qp} = \overline{p}\,\overline{q}$ for all $p, q \in \mathbb{H}$.

10. Prove that $\frac{1}{p} = \frac{\overline{p}}{|p|^2}$, where $p \in \mathbb{H} \setminus \{0_{\mathbb{H}}\}$.

11. Find the first matrix representation for the following quaternions: (a) $p = 1 - 2i + 3j + 4k$; (b) $p = 2 - i + j$; (c) $p = \frac{1}{2} + \frac{1}{3}i - 2j + k$.

12. Determine quaternions that correspond to the following matrices:

$$\text{(a)} \begin{pmatrix} 1 - i & 2 + 4i \\ -2 + 4i & 1 + i \end{pmatrix}; \quad \text{(b)} \begin{pmatrix} 1 & -2 + i \\ 2 + i & 1 \end{pmatrix}; \quad \text{(c)} \begin{pmatrix} 1 + \frac{1}{2}i & 2 + 8i \\ -2 + 8i & 1 - \frac{1}{2}i \end{pmatrix}.$$

13. Determine which of the following matrices correspond to a quaternion:

$$\text{(a)} \begin{pmatrix} 1 + i & 1 - i \\ 1 + i & 2 + i \end{pmatrix}; \quad \text{(b)} \begin{pmatrix} 2 + \frac{1}{2}i & 3 - i \\ -3 + i & 2 - \frac{1}{2}i \end{pmatrix}; \quad \text{(c)} \begin{pmatrix} 1 + 4i & 5 + 7i \\ 1 - i & 6 - 8i \end{pmatrix};$$

$$\text{(d)} \begin{pmatrix} 1 - 2i & 2 - i \\ -2 - i & 1 + 2i \end{pmatrix}.$$

14. Let $p = a + bi + cj + dk$, where $(a, b, c, d) \neq (0, 0, 0, 0)$, $a, b, c, d \in \mathbb{R}$. Let A be the first matrix representation of the quaternion p. Prove that $|p| = \sqrt{\det A}$.

15. Find the following products using the first matrix representation: (a) pq, $p = 1 - i + 2j - k$, $q = -2i + 3j + k$; (b) pq, $p = 2 - i - j$, $q = 1 + i + k$.

16. Using the first matrix representation compute $2p - 3q + pq$, where $p = 1 + i - j + k$ and $q = 2 - j$.

17. Write the second matrix representation of the following quaternions: (a) $p = 1 - 2i + \frac{1}{2}j + k$; (b) $p = i - j + 3k$; (c) $p = 2 + 4i - j - k$; (d) $p = i + j + k$.

18. Using the second matrix representation find pq, where $p = 1 - i + j + k$ and $q = 2 - i - j - k$.

19. Find the quaternions which correspond to the following matrices:

(a) $\begin{pmatrix} \frac{1}{2} & 2 & -9 & 3 \\ -2 & \frac{1}{2} & 3 & 9 \\ 9 & -3 & \frac{1}{2} & 2 \\ -3 & -9 & -2 & \frac{1}{2} \end{pmatrix}$; (b) $\begin{pmatrix} 7 & -1 & 9 & 0 \\ 1 & 7 & 0 & -9 \\ -9 & 0 & 7 & -1 \\ 0 & 9 & 1 & 0 \end{pmatrix}$;

(c) $\begin{pmatrix} 1 & 1 & 1 & -1 \\ -1 & 1 & -1 & -1 \\ -1 & 1 & 1 & 1 \\ 1 & 1 & -1 & 1 \end{pmatrix}$; (d) $\begin{pmatrix} 2 & 2 & -1 & -1 \\ -2 & 2 & 1 & 1 \\ 1 & -1 & 2 & 2 \\ -1 & -1 & -2 & 2 \end{pmatrix}$;

(e) $\begin{pmatrix} 3 & \frac{1}{2} & 1 & -1 \\ -\frac{1}{2} & 3 & -1 & -1 \\ -1 & 1 & 3 & \frac{1}{2} \\ 1 & 1 & -\frac{1}{2} & 3 \end{pmatrix}$; (f) $\begin{pmatrix} 1 & \frac{1}{3} & 4 & -\frac{2}{3} \\ -\frac{1}{3} & 1 & -\frac{2}{3} & -4 \\ -4 & \frac{2}{3} & 1 & \frac{1}{3} \\ \frac{2}{3} & 4 & -\frac{1}{3} & 1 \end{pmatrix}$.

20. Determine which of the following matrices correspond to a quaternion:

(a) $\begin{pmatrix} 1 & 0 & 2 & 3 \\ 2 & 1 & -1 & 4 \\ 3 & 2 & 1 & 1 \\ 1 & 1 & 1 & 1 \end{pmatrix}$; (b) $\begin{pmatrix} 0 & 0 & 0 & 0 \\ 2 & 1 & 1 & 1 \\ -1 & 1 & 1 & 1 \\ -1 & 0 & 2 & 0 \end{pmatrix}$;

(c) $\begin{pmatrix} 3 & -2 & 1 & -1 \\ 2 & 3 & -1 & -1 \\ -1 & 1 & 3 & -2 \\ 1 & 1 & 2 & 3 \end{pmatrix}$; (d) $\begin{pmatrix} 0 & 1 & 2 & 3 \\ 2 & 2 & 1 & 0 \\ 0 & 3 & 2 & 1 \\ 3 & 0 & 2 & 1 \end{pmatrix}$;

(e) $\begin{pmatrix} -1 & -1 & -1 & -1 \\ 1 & 0 & 0 & 0 \\ 2 & 0 & 1 & 1 \\ 1 & 1 & 1 & 1 \end{pmatrix}$; (f) $\begin{pmatrix} 2 & 1 & 0 & 0 \\ 2 & 3 & 0 & 1 \\ 1 & 1 & 1 & 2 \\ 3 & 2 & 1 & 1 \end{pmatrix}$.

21. Using the second matrix representation compute, $p - q\left(\frac{1}{2}p + 2q\right)$, where $p = 2 - 4i + 4j + 4k$ and $q = i - j - k$.

22. Compute the dot product $p \cdot q$, where: (a) $p = q = i$;   (b) $p = q = j$; (c) $p = q = k$;   (d) $p = i$, $q = j$;   (e) $p = i$, $q = k$;   (f) $p = j$, $q = i$;   (g) $p = j$, $q = k$;   (h) $p = k$, $q = i$;   (i) $p = k$, $q = j$; (j) $p = 1 - i + j + 2k$, $q = i + 3j - \frac{4}{3}k$;   (k) $p = i + j + k$, $q = 2 - i - j - k$; (l) $p = \frac{1}{2} + i + j$, $q = j + k$.

23. Compute $pq - p(p \cdot q + 2q) + q \cdot (p - q)$, where: (a) $p = 1 - i - j - k$, $q = 2 + i + j$;   (b) $p = i + j + k$, $q = 3 - i - j - k$;   (c) $p = 1 + i + j + k$, $q = 2 - i - k$.

24. Let $p = a_1 + b_1 i + c_1 j + d_1 k$ and $q = a_2 + b_2 i + c_2 j + d_2 k$, where $a_i, b_i, c_i, d_i \in \mathbb{R}$ ($i = 1, 2$). Compute $p \cdot q$.

25. Let $p = a_1 + b_1 i + c_1 j + d_1 k$ and $q = a_2 + b_2 i + c_2 j + d_2 k$, where $a_i, b_i, c_i, d_i \in \mathbb{R}$ ($i = 1, 2$). Prove that $p \cdot q = \frac{1}{2}(\overline{p}q + \overline{q}p)$.

26. Using the previous problem compute $p \cdot q$ if (a) $p = 1 - i + j$, $q = i - j - k$; (b) $p = 2 + i + j + k$, $q = j - k$;   (c) $p = 1 - i + k$, $q = i - j - k$.

27. Let $p = a + bi + cj + dk$, where $a, b, c, d \in \mathbb{R}$. Prove that $1_{\mathbb{H}} \cdot p = p_0$.

28. Compute the outer product (p, q) if (a) $p = 1 - i$, $q = j + k$;   (b) $p = 1 + i + k$, $q = 1 - i - j - k$;   (c) $p = j + k$, $q = i + k$.

29. Compute: (a) $p \cdot q - 2p + q - pq + q \cdot (p, q)$;   (b) $p - q + 2q \cdot p - q(p + q) - p \cdot (p, q)$ if $p = 1 - i - j - k$ and $q = 1 + i + j - k$.

30. Let $p = a_1 + b_1 i + c_1 j + d_1 k$ and $q = a_2 + b_2 i + c_2 j + d_2 k$, where $a_i, b_i, c_i, d_i \in \mathbb{R}$ ($i = 1, 2$). Find (p, q).

31. Let $p = a_1 + b_1 i + c_1 j + d_1 k$ and $q = a_2 + b_2 i + c_2 j + d_2 k$, where $a_i, b_i, c_i, d_i \in \mathbb{R}$ ($i = 1, 2$). Prove that $(p, q) = \frac{1}{2}(\overline{p}q - \overline{q}p)$.

32. Using the previous exercise, compute (p, q) if (a) $p = i + j$, $q = i + j + k$; (b) $p = -2 - 2i - 3j + k$, $q = 1 - i - k$;   (c) $p = -1 - i - 2k$, $q = 2 - 3i + 4j$;   (d) $p = 2 - i$, $q = 3 + k$;   (e) $p = 3 + i + j$, $q = i + k$;   (f) $p = 1 - i - j$, $q = 2 - i - k$;   (g) $p = -i$, $q = j$; (h) $p = i$, $q = k$;   (i) $p = j$, $q = k$.

33. Prove that $(p, q) = -(q, p)$, where $p, q \in \mathbb{H}$.

34. Prove that $(p, q) = \mathrm{Vec}(\overline{p}q) = -\mathrm{Vec}(\overline{q}p)$, where $p, q \in \mathbb{H}$.

35. Compute $[p, q]$, where (a) $p = 2 - i + j - k$, $q = 1 + i + j + k$;   (b) $p = 3 + j$, $q = 2 + k$;   (c) $p = 1 - 2i + 2j + k$, $q = 1 + i + 2j + k$;   (d) $p = i$, $q = j$;   (e) $p = i$, $q = k$;   (f) $p = j$, $q = k$.

36. Compute $2p + q - pq + q \cdot ([p, q] + p) + (p, q)$, where $p = 1 + i + j - k$ and $q = 2 - i - j + k$.

37. Let $p = a_1 + b_1 i + c_1 j + d_1 k$ and $q = a_2 + b_2 i + c_2 j + d_2 k$, where $a_i, b_i, c_i, d_i \in \mathbb{R}$ ($i = 1, 2$). Compute $[p, q]$.

38. Prove that $[p, q] = \frac{1}{2}(pq + qp)$, where p, q are real quaternions.

39. Using the previous exercise, compute $[p, q]$, where (a) $p = i + j$, $q = i - j$; (b) $p = 2 + i + j + k$, $q = i$;   (c) $p = 1 - i - j - k$, $q = i + j + k$.

40. Prove that $[p, q] = [q, p]$, where $p, q \in \mathbb{H}$.

41. Compute $p \times q$ if (a) $p = i + 2j - k$, $q = -i + j + k$;   (b) $p = 3k$, $q = -i - j + k$;   (c) $p = -\frac{1}{2}i + j + k$, $q = -i + j - k$;   (d) $p = i$, $q = j$;   (e) $p = i$, $q = k$;   (f) $p = j$, $q = k$.

42. Let $p = b_1 i + c_1 j + d_1 k$ and $q = b_2 i + c_2 j + d_2 k$, where $b_i, c_i, d_i \in \mathbb{R}$ $(i = 1, 2)$. Compute $p \times q$.

43. Prove that $p \times q = \frac{1}{2}(pq - qp)$, where $p, q \in \mathbb{H}$.

44. Using the previous exercise, compute $p \times q$ if (a) $p = i + j - k$, $q = -i + j + k$;   (b) $p = 2 - i + j + k$, $q = i - k$;   (c) $p = 3 - i + j + 2k$, $q = 2 - i + k$.

45. Compute (a) $p \cdot q - (p, q) + p \times q$;   (b) $[p, q] - 2p \times q + (p, q)$;   (c) $p \cdot (p + q) - p \times (p - q)$, if $p = 1 - i + j + k$ and $q = 2 - i - k$.

46. Compute (a) $2p - r - r \cdot (p + q) - p(q + r)$;   (b) $p - [q, q] + (p, r)$; (c) $p \times r + p \times q + q \times r - p \cdot q$;   (d) $p - [p, 2p - r] + q \times (p + r)$, where $p = i + j, q = 1 - i - k$ and $r = i + j + k$.

47. Find the sign of the following quaternions: (a) $p = 3 + i - j + k$;   (b) $p = 1 - i + k$;   (c) $p = i - j$.

48. Compute (a) $pq - p + 2q - 3\operatorname{sgn}(r)$;   (b) $[p, q] - 2(p, r) + (p \times r)\operatorname{sgn}(q)$; (c) $|p|(q - r) - (p, q)\operatorname{sgn}(p)$;   (d) $p^2 - q\operatorname{sgn}(r)$;   (e) $pqr - \operatorname{sgn}(p) - 3p$; (f) $p\operatorname{sgn}(q) - 3r$, where $p = i + j, q = 1 - k$ and $r = i + j + k$.

49. Compute $\operatorname{sgn}(p)$, where $p = a + bi + cj + dk \in \mathbb{H}$.

50. Find the arguments of the following quaternions: (a) $p = 1 + 2i - j$;   (b) $p = -1 + i + j - k$;   (c) $p = 2 - i + k$.

51. Compute (a) $\arg(q)(p - r) - [q, r]$;   (b) $(p, q) - \arg(r)p \cdot q + \operatorname{sgn}(r)$;   (c) $p^2 - q^2 - \arg(r)(p + q)$, where $p = 2 + j + k, q = 1 - i - k$ and $r = j + k$.

52. Find the trigonometric representations of the following quaternions: (a) $p = 1 + i + j + k$;   (b) $p = i + j$;   (c) $p = i + k$;   (d) $p = j + k$; (e) $p = i - j$;   (f) $p = i - k$;   (g) $p = j - k$.

53. Write the first decomposition of the following quaternions: (a) $p = 1 + i + j + k$;   (b) $p = 2 - \frac{3}{2}i - 4j + 5k$;   (c) $p = \frac{1}{2} - 2i + 3j - 4k$.

54. Compute pq using the first decomposition, where (a) $p = j + k$, $q = 1 - i - k$; (b) $p = 2 + \frac{3}{2}i - j - k$, $q = 1 - i - j - k$;   (c) $p = i + j + k$, $q = 2 - 3i + 4k$.

55. Write the first decomposition for the following quaternions:

$$\text{(a)} \begin{pmatrix} 1 & -2 & 3 & -4 \\ 2 & 1 & -4 & -3 \\ -3 & 4 & 1 & -2 \\ 4 & 3 & 2 & 1 \end{pmatrix}; \qquad \text{(b)} \begin{pmatrix} 1 & -1 & 1 & -1 \\ 1 & 1 & -1 & -1 \\ -1 & 1 & 1 & -1 \\ 1 & 1 & 1 & 1 \end{pmatrix};$$

$$\text{(c)} \begin{pmatrix} 1 - i & -2 + 4i \\ 2 + 4i & 1 + i \end{pmatrix}.$$

56. Find p_+ and p_-, where (a) $p = 2 + \frac{3}{2}i - \frac{1}{2}j + k$;   (b) $p = 1 + i - j + 2k$;   (c) $p = 1 - i + j - 2k$.

57. Let $p = a + bi + cj + dk$, where $a, b, c, d \in \mathbb{R}$. Find p_+ and p_-.

58. Let $p = a + bi + cj + dk$, where $a, b, c, d \in \mathbb{R}$. Prove that

$$p_+ = \left(a + d + i(b - c)\right)\frac{1+k}{2}, \qquad p_- = \left(a - d + i(b + c)\right)\frac{1-k}{2}.$$

59. Let $p = a + bi + cj + dk$, where $a, b, c, d \in \mathbb{R}$. Prove that

$$p_+ = \frac{1+k}{2}\left(a + d + j(c - b)\right), \qquad p_- = \frac{1-k}{2}\left(a - d + j(b + c)\right).$$

60. Prove that $|p|^2 = |p_-|^2 + |p_+|^2$.

61. Let $p, q \in \mathbb{H}$. Prove that $\mathrm{Sc}(p_+\overline{q}_-) = \mathrm{Sc}(p_-\overline{q}_+) = 0$.

62. Let $p = 1 - i - j - k$, and $q = j + k$. Compute (a) $p^2 + p_+^2 - p_-^2$;   (b) $p + p_+ - p_+p_-$;   (c) $p + p_+q + p_-(i + q)$;   (d) $q + p_+ \cdot p_-$;   (e) $p_+^3 - \overline{p_-}q$.

63. Check the relation $\mathrm{Sc}\,(qv\overline{q}) = 0$, where (a) $q = 1 - i + j + k$, $v = 1 + i + j + k$;   (b) $q = j$, $v = 10 - 2i - j + k$;   (c) $q = i + j + k$, $v = 1 - i - j - k$.

64. Let $p, q \in \mathbb{H}$. Prove that $\mathrm{Sc}\,(pq\overline{p}) = 0$.

65. Check the relation $|q\mathbf{p}\overline{q}| = |\mathbf{p}|$, where (a) $q = \frac{1}{2} - \frac{1}{2}i + \frac{1}{2}j + \frac{1}{2}k$, $p = i - j - k$;   (b) $q = \frac{1}{\sqrt{2}} - \frac{1}{\sqrt{2}}k$, $p = 2 - 3i + j - k$;   (c) $q = \frac{1}{\sqrt{3}} - \frac{1}{\sqrt{3}}j + \frac{1}{\sqrt{3}}k$, $p = 1 - i + 3j - k$.

66. Let q be a unit real quaternion, and $p \in \mathbb{H}$. Prove that $|q\mathbf{p}\overline{q}| = |\mathbf{p}|$.

67. Check the relation $q\,(\alpha\mathbf{p} + \mathbf{q})\,\overline{q} = \alpha q\mathbf{p}\overline{q} + q\mathbf{q}\overline{q}$, where (a) $\alpha = 3$, $q = 1 - i + j - k$, $p = i + j$, $q = i - j$;   (b) $\alpha = -2$, $q = i - j + k$, $p = 1 + i - j - k$, $q = i + k$;   (c) $\alpha = 1$, $q = 2 - \frac{1}{2}i + j - k$, $p = 1 - i + j - k$, $q = i - j + k$.

68. Let $\alpha \in \mathbb{R}$, and $p, q \in \mathbb{H}$. Prove that $q\,(\alpha\mathbf{p} + \mathbf{q})\,\overline{q} = \alpha q\mathbf{p}\overline{q} + q\mathbf{q}\overline{q}$.

69. Let q be a unit real quaternion, $p \in \mathbb{H}$ and $R(p) = q\mathbf{p}\overline{q}$. Prove that $R(p)$ is an orthonormal transformation.

70. Compute $p\mathbf{p}\overline{p}$, where (a) $p = 1 - \frac{1}{\sqrt{3}}i + \frac{1}{\sqrt{3}}j - \frac{1}{\sqrt{3}}k$;   (b) $p = i$;   (c) $p = j$;   (d) $p = k$;   (e) $p = 10 - \frac{1}{6}i + \frac{2}{\sqrt{6}}j - \frac{1}{\sqrt{6}}k$;   (f) $p = 1 - \frac{1}{\sqrt{19}}i + \frac{3}{\sqrt{19}}j - \frac{3}{\sqrt{19}}k$.

71. Let $p \in \mathbb{H}$ such that $|\mathbf{p}| = 1$. Prove that (a) $\mathbf{p}^2 = -1_{\mathbb{H}}$;   (b) $\mathbf{p}^3 = -\mathbf{p}$;   (c) $\left(p_0^2 + 1_{\mathbb{H}}\right)\mathbf{p} = p\mathbf{p}\overline{p}$.

Quaternions and Spatial Rotation **2**

The particularly rich theory of rotations does not need advertising. One can think of a rotation as a transformation in the plane or in space that describes the position and orientation of a three-dimensional rigid body around a fixed point. The first ever study of rotations was published by L. Euler in 1776. Since then the analysis of 3D rotations has been the object of intensive study in many fields such as, for example, computer graphics, computer vision, robotics, navigation, molecular dynamics and orbital mechanics of satellites. As the reader will soon discover, unit quaternions are very efficient for analyzing situations in which rotations in $\mathbb{R}^3$ are involved. Due to the buildup of computational errors, the quaternion might become of non-unit length. Nevertheless, this can be fixed by dividing the quaternion by its norm. In contrast to the Euler angles, unit quaternions are very simple to compose and avoid the gimbal lock problem (discontinuities). Besides this, compared to rotation matrices they are numerically much more stable and may be more efficient. For these reasons, quaternions are commonly used in computer graphics and navigation systems.

In this chapter we shall briefly discuss the close connection between quaternions and rotations. This is exactly what this part will try to examine. It does not add any application to rotation interpolation (for which excellent surveys already exist) but gives an overview of the problems which thus arise. The underlying quaternion technique will also allow us to easily compose rotations and to understand their geometric meaning. The rotations that we will discuss here are rotations about the origin, a fitting point to the present exposition.

To start with, each unit vector is to be considered as an axis of rotation and the quaternion that represents a rotation about that axis is to be associated with the vector. To sidestep this issue we shall understand how to turn a rotation about one axis into a rotation about another axis. The key point is a similarity transformation, which we now define:

J.P. Morais et al., *Real Quaternionic Calculus Handbook*,
DOI 10.1007/978-3-0348-0622-0_2, © Springer Basel 2014

2.1 Rotations

Let p be an arbitrary quaternion. For every quaternion $q \neq 0_{\mathbb{H}}$, the mapping ρ_q in $\mathbb{H}$ defined by $\rho_q(p) := qpq^{-1}$ is called rotation. We have then the following properties:

(i) $\mathbb{R}$-linearity
$$\rho_q(\lambda_1 p_1 + \lambda_2 p_2) = \lambda_1 \rho_q(p_1) + \lambda_2 \rho_q(p_2) \quad \text{for all } p_1, p_2 \in \mathbb{H},\ \lambda_1, \lambda_2 \in \mathbb{R};$$
(ii) Multiplicativity
$$\rho_q(p_1 p_2) = \rho_q(p_1)\rho_q(p_2) \quad \text{for all } p_1, p_2 \in \mathbb{H};$$
(iii) ρ_q is an isometric automorphism of $\mathbb{H}$;
(iv) $\rho_q(p_1) \cdot \rho_q(p_2) = p_1 \cdot p_2 \quad \text{for all } p_1, p_2 \in \mathbb{H};$
(v) $\rho_{q_1}\rho_{q_2}(p) = \rho_{q_1 q_2}(p) \quad \text{for all } q_1, q_2 \in \mathbb{H} \setminus \{0_{\mathbb{H}}\}.$

For simplicity we only sketch the proofs of Properties (iii)–(v). A detailed verification of the remaining statements is left to the reader. Obviously, $\rho_q^{-1} : p \mapsto q^{-1}pq$, and ρ_q^{-1} corresponds to the inverse of ρ_q; also, $\rho_q^{-1} = \rho_{q^{-1}}$. Together with (ii) it follows that ρ_q is an automorphism. Moreover, since

$$|\rho_q(p)| = |qpq^{-1}| = |q||p||q^{-1}|$$

$$= |q||p|\left|\frac{\overline{q}}{|q|^2}\right| = |p|,$$

(iii) is proved. For the proof of (iv), a direct computation shows that

$$\rho_q(p_1) \cdot \rho_q(p_2) = \frac{1}{2}\left[\rho_q(p_2)\overline{\rho_q(p_1)} + \rho_q(p_1)\overline{\rho_q(p_2)}\right]$$

$$= \frac{1}{2}\left[qp_2 q^{-1}\overline{qp_1 q^{-1}} + qp_1 q^{-1}\overline{qp_2 q^{-1}}\right]$$

$$= \frac{1}{2}\left[qp_2 q^{-1}\overline{q^{-1}}\,\overline{p_1}\,\overline{q} + qp_1 q^{-1}\overline{q^{-1}}\,\overline{p_2}\,\overline{q}\right]$$

$$= \frac{1}{2}\left[q(p_2\overline{p_1} + p_1\overline{p_2})\overline{q}\,\frac{1}{|q|^2}\right]$$

$$= p_1 \cdot p_2.$$

The associativity yields the relation

$$\rho_{q_1}\rho_{q_2}(p) = q_1(q_2 p q_2^{-1})q_1^{-1} = q_1 q_2 p (q_1 q_2)^{-1} = \rho_{q_1 q_2}(p).$$

Example. Let $p = 1 + 101i + 203j + 401k$, $q_1 = 1 + i + j + k$, and $q_2 = \frac{1}{2}(1 - i - j - k)$. Find $\rho_{q_1}\rho_{q_2}(p)$.

Solution. A direct calculation shows that

$$\rho_{q_1}\rho_{q_2}(p) = (q_1 q_2)p(q_1 q_2)^{-1} = 2(1 + 101i + 203j + 401k)\frac{1}{2} = p.$$

Exercise 2.54. *Let* $p_1 = -i + j + k$, $p_2 = 2i + j$, *and* $q = 1 + j$. *Find* $\rho_q(p_1) \times \rho_q(p_2)$.

Solution. $-\sqrt{2}(3i + 2j - k)$.

Exercise 2.55. *Let* $p_1 = 3i - 5k$, $p_2 = i - j$, *and* $q = i - 2j$. *Compute* (a) $(\rho_q(p_1), \rho_q(p_2))$;   (b) $[\rho_q(p_1), \rho_q(p_2)]$.

Solution. (a) $5i + j - 3k$;   (b) $-3 + 5i + j + 3k$.

Exercise 2.56. *Prove that the direction of* p, *if along* q, *is left unchanged by the automorphism* $\rho_q(p)$.

Exercise 2.57. *Let* p *and* q *be quaternions. Prove that* ρ_q *does not depend on* $|q|$.

Exercise 2.58. *Let* p, q_1 *and* q_2 *be quaternions such that* $q_1, q_2 \neq 0_{\mathbb{H}}$. *Show that* $\rho_{q_1}(p)$ *and* $\rho_{q_2}(p)$ *yield the same rotation if and only if* $q_1 = \lambda q_2$ *for some real number* $\lambda \neq 0$.

The next example provides an alternative way of computing $\rho_q(p)$.

Example. Let p and q be quaternions such that $q \neq 0_{\mathbb{H}}$. Prove the following identity:

$$\rho_q(p) = p_0 + \text{Vec}(\rho_q(p)), \tag{2.1}$$

where

$$\text{Vec}(\rho_q(p)) = \frac{1}{|q|^2}\left[(q_0^2 - \mathbf{q} \cdot \mathbf{q})\mathbf{p} + 2(\mathbf{p} \cdot \mathbf{q})\mathbf{q} - 2q_0(\mathbf{p} \times \mathbf{q})\right].$$

Solution. First, a straightforward observation is that $qpq^{-1} = qp_0q^{-1} + q\mathbf{p}q^{-1}$. It follows that $qp_0q^{-1} = p_0$, while

$$\text{Sc}(q\mathbf{p}q^{-1}) = \frac{1}{|q|^2}\left[q_0(\mathbf{p} \cdot \mathbf{q}) - \mathbf{q} \cdot (q_0\mathbf{p} - (\mathbf{p} \times \mathbf{q}))\right] = 0$$

and

$$\text{Vec}(q\mathbf{p}q^{-1}) = \frac{1}{|q|^2}\left[q_0^2\mathbf{p} - q_0(\mathbf{p} \times \mathbf{q}) + (\mathbf{p} \cdot \mathbf{q})\mathbf{q} + q_0(\mathbf{q} \times \mathbf{p}) - \mathbf{q} \times (\mathbf{p} \times \mathbf{q})\right].$$

For three pure quaternions $\mathbf{n}_1$, $\mathbf{n}_2$, $\mathbf{n}_3$, it holds

$$\mathbf{n}_1 \times (\mathbf{n}_2 \times \mathbf{n}_3) = (\mathbf{n}_1 \cdot \mathbf{n}_3)\mathbf{n}_2 - (\mathbf{n}_1 \cdot \mathbf{n}_2)\mathbf{n}_3,$$

whence

$$\rho_q(p) = p_0 + \frac{1}{|q|^2}\left[(q_0^2 - \mathbf{q}\cdot\mathbf{q})\mathbf{p} + 2(\mathbf{p}\cdot\mathbf{q})\mathbf{q} - 2q_0(\mathbf{p}\times\mathbf{q})\right].$$

Suppose that we restrict ourselves to purely imaginary quaternions $\mathbf{p}$ in the mapping ρ_q. Let us emphasize that $\rho_q(\mathbf{p})$ is again a pure quaternion, so that the scalar part of $\rho_q(\mathbf{p})$ does not exist. Thus $\rho_q(\mathbf{p})$ is also an automorphism of $\mathbb{R}^3$ with the properties described above. Furthermore, $\rho_q(\mathbf{p})$ in $\mathbb{R}^3$ is invariant under the vector product, namely

$$\begin{aligned}
\rho_q(\mathbf{p}_1) \times \rho_q(\mathbf{p}_2) &= \frac{1}{2}\left[\rho_q(\mathbf{p}_1)\rho_q(\mathbf{p}_2) - \rho_q(\mathbf{p}_2)\rho_q(\mathbf{p}_1)\right] \\
&= \frac{1}{2}\left[q\mathbf{p}_1 q^{-1}q\mathbf{p}_2 q^{-1} - q\mathbf{p}_2 q^{-1}q\mathbf{p}_1 q^{-1}\right] \\
&= \frac{1}{2}q\left[\mathbf{p}_1\mathbf{p}_2 - \mathbf{p}_2\mathbf{p}_1\right]q^{-1} \\
&= \rho_q(\mathbf{p}_1 \times \mathbf{p}_2).
\end{aligned}$$

In summary, the mapping ρ_q is an automorphism of $\mathbb{R}^3$ that leaves the canonical scalar product invariant. Following the above calculations, it is homomorphic with respect to the vector product.

Exercise 2.59. *Let $\mathbf{p}_1 = 2i + j + k$, $\mathbf{p}_2 = j - k$, and $q = j$. Find $\rho_q(\mathbf{p}_1) \times \rho_q(\mathbf{p}_2)$.*

Solution. $2i + 2j - 2k$.

Exercise 2.60. *Let $p = 1 + 2i + 2j - k$. Prove that $\rho_p(p_+) = 0_{\mathbb{H}}$.*

Exercise 2.61. *Let $p = a + bi + cj + dk$. Under what circumstances does $\rho_{\mathbf{p}_-}(\mathbf{p}_+) = 0_{\mathbb{H}}$ hold?*

Solution. $b = c \neq 0$, $d = -a \neq 0$.

Before we proceed further, let us give an example:

Example. Let q be a unit quaternion and $\mathbf{p}$ a pure quaternion. Prove that $\frac{\mathbf{q}}{|\mathbf{q}|}$ remains invariant under the mapping ρ_q.

Solution. Let $q = q_0 + \mathbf{q}$ with $q_0^2 + |\mathbf{q}|^2 = 1$. A first straightforward computation shows that

$$
\rho_q\left(\frac{\mathbf{q}}{|\mathbf{q}|}\right) = \left(q_0 + \frac{\mathbf{q}}{|\mathbf{q}|}|\mathbf{q}|\right)\frac{\mathbf{q}}{|\mathbf{q}|}\left(q_0 - \frac{\mathbf{q}}{|\mathbf{q}|}|\mathbf{q}|\right)
$$

$$
= \left(q_0 + \frac{\mathbf{q}}{|\mathbf{q}|}|\mathbf{q}|\right)\left(q_0 - \frac{\mathbf{q}}{|\mathbf{q}|}|\mathbf{q}|\right)\frac{\mathbf{q}}{|\mathbf{q}|}
$$

$$
= (q_0^2 + |\mathbf{q}|^2)\frac{\mathbf{q}}{|\mathbf{q}|}
$$

$$
= \frac{\mathbf{q}}{|\mathbf{q}|}.
$$

2.2   Composition of Rotations

From a computational point of view, the quaternion representation of rotation simplifies the computation of composite rotations. Consider two rotations ρ_{q_1} and ρ_{q_2} specified by two unit quaternions q_1 and q_2. The action of the composite rotation $\rho_{q_2} \circ \rho_{q_1}$ on a pure quaternion $\mathbf{p}$, which is obtained by first applying the rotation induced by q_1 and then applying the rotation induced by q_2, is the same as applying the single rotation induced by the product $q_2 q_1$, namely

$$
(\rho_{q_2} \circ \rho_{q_1})(\mathbf{p}) = \rho_{q_2}(\rho_{q_1}(\mathbf{p}))
$$

$$
= q_2(q_1 \mathbf{p} q_1^{-1})q_2^{-1} = (q_2 q_1)\mathbf{p}(q_1^{-1}q_2^{-1})
$$

$$
= (q_2 q_1)\mathbf{p}(q_2 q_1)^{-1} = \rho_{q_2 q_1}(\mathbf{p}).
$$

Since q_1 and q_2 are unit quaternions, it is evident that the product $q_2 q_1$ is also a unit quaternion. Accordingly, $\rho_{q_2 q_1}(\mathbf{p})$ describes a rotation operator whose defining quaternion is the product of the two quaternions q_1 and q_2. Thus the axis and the angle of the composite rotation are given by the product $q_2 q_1$. Similarly, the quaternion product $q_1 q_2$ defines an operator $\rho_{(q_1 q_2)^{-1}}$, which represents the composition of the operator $\rho_{q_1^{-1}}$ followed by $\rho_{q_2^{-1}}$. The axis and angle of rotation of $\rho_{(q_1 q_2)^{-1}}$ are those represented by the quaternion product $q_1 q_2$. In addition, we note that these considerations can be extended to the composition of any finite number of rotations.

Exercise 2.62. *Let $q_1 = i$, $q_2 = j$ and $\mathbf{p} = i + j$. Find $(\rho_{q_2} \circ \rho_{q_1})(\mathbf{p})$.*

Solution. $i + j$.

Exercise 2.63. *Let q_1, q_2 and q_3 be unit quaternions. What rotation is represented by the quaternion product $q_1 q_3 q_2$? Answer by specifying its rotation angle and axis.*

Solution. It is a composition of tree rotations, $\rho_{q_1} \circ \rho_{q_3} \circ \rho_{q_2}$.

We make the following geometric observations concerning a quaternion q, its inverse q^{-1} and its inverse with respect to addition $-q$:
  (i) *The inverse of q, q^{-1}, reverses the direction of rotation. Notice, in particular, that $\rho_{q^{-1}}\left(\rho_q(p)\right) = q^{-1}(qpq^{-1})q = p$;*
 (ii) *It can be easily seen that $\rho_{-q}(p) = (-q)p(-q)^{-1} = \rho_q(p)$. Therefore the quaternion $-q$ represents exactly the same rotation as q.*

Exercise 2.64. *Let $q_1 = 1 - i - j - k$, $q_2 = \frac{1}{2}(1 + i + j + k)$, and $\mathbf{p} = 101i - 1{,}031j - 76k$. Find $\rho_{q_1}\left(\rho_{q_2}(\mathbf{p})\right)$ and $\rho_{(q_1 q_2)^{-1}}(\mathbf{p})$.*

Solution. $\rho_{q_1}\left(\rho_{q_2}(\mathbf{p})\right) = \rho_{(q_1 q_2)^{-1}}(\mathbf{p}) = 101i - 1{,}031j - 76k.$

Next we address the issue of computing the matrix representation of a rotation.

2.3 Rotation Matrix Representation

Let $p := p_0 + p_1 i + p_2 j + p_3 k$ and $q := a + bi + cj + dk \neq 0_{\mathbb{H}}$. After some calculations, we find that

$$
\rho_q(p)|q|^2 = (a^2 + b^2 + c^2 + d^2)p_0
$$

$$
+ \left[(a^2 + b^2 - c^2 - d^2)p_1 + (-2ad + 2bc)p_2 + (2ac + 2bd)p_3\right] i
$$

$$
+ \left[(2ad + 2bc)p_1 + (a^2 - b^2 + c^2 - d^2)p_2 + (-2ab + 2cd)p_3\right] j
$$

$$
+ \left[(-2ac + 2bd)p_1 + (2ab + 2cd)p_2 + (a^2 - b^2 - c^2 + d^2)p_3\right] k.
$$

If we identify a quaternion $p \in \mathbb{H}$ with a column vector in $\mathbb{R}^4$, we can write

$$
\rho_q(p) \leftrightarrow \frac{1}{|q|^2}
\begin{pmatrix}
(a^2 + b^2 + c^2 + d^2)p_0 \\
(a^2 + b^2 - c^2 - d^2)p_1 + (-2ad + 2bc)p_2 + (2ac + 2bd)p_3 \\
(2ad + 2bc)p_1 + (a^2 - b^2 + c^2 - d^2)p_2 + (-2ab + 2cd)p_3 \\
(-2ac + 2bd)p_1 + (2ab + 2cd)p_2 + (a^2 - b^2 - c^2 + d^2)p_3
\end{pmatrix}.
$$

The matrix that acts upon the 4D-column vectors $(p_0\ \ p_1\ \ p_2\ \ p_3)^T$,

$$
\mathcal{M}(\rho_q) := (\mathcal{M}_{i,j})_{i,j=1,\ldots,4},
$$

is explicitly defined as

$$\frac{1}{|q|^2}\begin{pmatrix} |q|^2 & 0 & 0 & 0 \\ 0 & a^2+b^2-c^2-d^2 & 2(bc-ad) & 2(ac+bd) \\ 0 & 2(ad+bc) & a^2-b^2+c^2-d^2 & 2(cd-ab) \\ 0 & 2(bd-ac) & 2(ab+cd) & a^2-b^2-c^2+d^2 \end{pmatrix}.$$

The matrix $\mathcal{M}(\rho_q)$ has the following properties:

(i)   The dot product of any row with itself is $|q|^4 = (a^2+b^2+c^2+d^2)^2$. Similarly for the dot product of any column with itself;

(ii) The dot product of any row with any other row is zero. Similarly for any column with any other column;

(iii) $\det\left(\mathcal{M}(\rho_q)\right) > 0$ for all $q \neq 0_{\mathbb{H}}$;

(iv) If q_1 and q_2 are any two quaternions, $\mathcal{M}(\rho_{q_1})\mathcal{M}(\rho_{q_2}) = \mathcal{M}(\rho_{q_1 q_2})$, where the multiplication on the left-hand side corresponds to matrix multiplication and the one on the right to quaternion multiplication.

The verification of these statements is not difficult and is left to the reader. It can be easily seen that $\mathcal{M}(\rho_q)$ is an orthogonal matrix. Hence $\mathcal{M}(\rho_q)$ describes a rotation about the origin in four-dimensional Euclidean space. On noting that $\det\left(\mathcal{M}(\rho_q)\right)$ is positive we find that $\mathcal{M}(\rho_q)$ describes a proper rotation. Besides preserving length, $\mathcal{M}(\rho_q)$ preserves orientation as well.

In particular, from the diagonal elements of $\mathcal{M}(\rho_q)$ we get

$$a^2 = \frac{|q|^2}{4}\left[\mathcal{M}_{1,1}(\rho_q) + \mathcal{M}_{2,2}(\rho_q) + \mathcal{M}_{3,3}(\rho_q) + \mathcal{M}_{4,4}(\rho_q)\right],$$

$$b^2 = \frac{|q|^2}{4}\left[\mathcal{M}_{1,1}(\rho_q) + \mathcal{M}_{2,2}(\rho_q) - \mathcal{M}_{3,3}(\rho_q) - \mathcal{M}_{4,4}(\rho_q)\right],$$

$$c^2 = \frac{|q|^2}{4}\left[\mathcal{M}_{1,1}(\rho_q) - \mathcal{M}_{2,2}(\rho_q) + \mathcal{M}_{3,3}(\rho_q) - \mathcal{M}_{4,4}(\rho_q)\right],$$

$$d^2 = \frac{|q|^2}{4}\left[\mathcal{M}_{1,1}(\rho_q) - \mathcal{M}_{2,2}(\rho_q) - \mathcal{M}_{3,3}(\rho_q) + \mathcal{M}_{4,4}(\rho_q)\right].$$

Besides these immediate identities there are also the following relevant relations:

$$ab = \frac{\mathcal{M}_{4,3}(\rho_q) - \mathcal{M}_{3,4}(\rho_q)}{4},$$

$$ac = \frac{\mathcal{M}_{2,4}(\rho_q) - \mathcal{M}_{4,2}(\rho_q)}{4} = \frac{\mathcal{M}_{3,2}(\rho_q) - \mathcal{M}_{2,3}(\rho_q)}{4},$$

$$bc = \frac{\mathcal{M}_{2,3}(\rho_q) + \mathcal{M}_{3,2}(\rho_q)}{4}, \qquad bd = \frac{\mathcal{M}_{2,4}(\rho_q) + \mathcal{M}_{4,2}(\rho_q)}{4},$$

$$cd = \frac{\mathcal{M}_{3,4}(\rho_q) + \mathcal{M}_{4,3}(\rho_q)}{4}.$$

Exercise 2.65. *Let $\mathcal{M}(\rho_q)$ be as in the previous exercise, and let $q = 1 - i - j - k$. Find* (a) $\det(\mathcal{M}(\rho_q))$;   (b) $\mathcal{M}_{1,3}(\rho_q)$;   (c) $\mathcal{M}_{3,2}(\rho_q)$.

Solution. (a) 1;   (b) 0;   (c) 0.

Exercise 2.66. *Let* $\mathcal{M}(\rho_q) = \begin{pmatrix} 1 & 0 & 0 & 0 \\ 0 & 0 & 1 & 0 \\ 0 & 0 & 0 & 1 \\ 0 & 1 & 0 & 0 \end{pmatrix}$. *Find* $q \neq 0_{\mathbb{H}}$.

Solution. $q = 1 - i - j - k$.

Exercise 2.67. *Let* $q = 1 + i + j + k$. *Find* $\mathcal{M}(\rho_q)$.

Solution. $\begin{pmatrix} 1 & 0 & 0 & 0 \\ 0 & 0 & 0 & 1 \\ 0 & 1 & 0 & 0 \\ 0 & 0 & 1 & 0 \end{pmatrix}$.

Suppose that we restrict ourselves to the action of the mapping ρ_q on purely imaginary quaternions $\mathbf{p}$. As we will see, for any pure quaternion $\mathbf{p}$ the action of ρ_q on $\mathbf{p}$ is equivalent to a rotation of $\mathbf{p}$ through an angle θ about the axis of rotation q. To explore this further, we need to study the properties of the automorphism ρ_q in $\mathbb{R}^3$.

Next we discuss the effect of a rotation through an angle θ about an axis passing through the origin.

2.4   Quaternion Representation of Rotations

Let p be a pure quaternion. In case q is a unit quaternion, $\rho_q(p)$ represents a counterclockwise rotation by the angle $2 \arccos(q_0)$ (measured between 0 and 2π) with axis pointing in the direction of $\mathbf{q}$. We set

$$q_0 := \cos\frac{\theta}{2} \geq 0, \quad \text{and} \quad \mathbf{q} := \frac{\mathbf{p}}{|\mathbf{p}|}\sin\frac{\theta}{2}.$$

If $\mathbf{q} = (0, 0, 0)$, then $q = \pm 1_{\mathbb{H}}$, which gives rise to the identity rotation (θ is either 0 or π). Therefore, we may assume that $|\mathbf{q}| \neq 0$. Any unit quaternion q can be written as

$$q = \cos\frac{\theta}{2} + \frac{\mathbf{p}}{|\mathbf{p}|}\sin\frac{\theta}{2}, \tag{2.2}$$

where the angle of rotation θ is equal to π when $q_0 = 0$, and when $q_0 \neq 0$, choosing a suitable orientation of the plane orthogonal to the axis of rotation, is given by $\tan\frac{\theta}{2} = \frac{|\mathbf{q}|}{q_0}$, with $0 < \theta \leq \pi$. It is evident that $\mathbf{q}$ gives the direction of the axis of rotation, and θ can be recovered from the scalar part and the magnitude of the vector part. The underlying two conditions,

$$q_0 = \cos\frac{\theta}{2}, \quad \text{and} \quad |\mathbf{q}|^2 = \sin^2\frac{\theta}{2},$$

are consistent since, obviously we have $|q|^2 = q_0^2 + |\mathbf{q}|^2 = 1$. We can add any multiple of $360°$ to θ without affecting the rotation. Notice, in particular, that the effect of a rotation through an angle θ in the opposite direction can be calculated by replacing θ by $-\theta$ in the representation (2.2). One can check directly that

$$\rho_q(p) = (q_0 + \mathbf{q})p(q_0 - \mathbf{q}) = q_0^2 p + q_0(\mathbf{q}p - p\mathbf{q}) - \mathbf{q}p\mathbf{q}. \tag{2.3}$$

In particular,

$$\rho_q(\mathbf{q}) = q_0^2\mathbf{q} + |\mathbf{q}|^2\mathbf{q} = \mathbf{q}.$$

Now we choose a pure unit quaternion $\mathbf{p}_1$ perpendicular to $\mathbf{q}$. Hence,

$$\mathbf{q}\mathbf{p}_1 = -\mathbf{p}_1\mathbf{q} = \mathbf{q} \times \mathbf{p}_1.$$

If we set

$$\mathbf{p}_3 = \frac{\mathbf{q}}{|\mathbf{q}|}, \qquad \mathbf{p}_2 = \mathbf{p}_3\mathbf{p}_1,$$

then $(\mathbf{p}_1, \mathbf{p}_2, \mathbf{p}_3)$ is an orthonormal basis of $\mathbb{R}^3$ with $\mathbf{p}_1 \times \mathbf{p}_2 = \mathbf{p}_3$. Relying on (2.3), direct computations show that

$$\rho_q(\mathbf{p}_1) = \mathbf{p}_1 \cos^2\frac{\theta}{2} + 2\mathbf{q}\mathbf{p}_1 \cos\frac{\theta}{2} - \mathbf{p}_1 \sin^2\frac{\theta}{2}$$

$$= \mathbf{p}_1 \cos\theta + \mathbf{p}_2 \sin\theta.$$

Similarly, one has

$$\rho_q(\mathbf{p}_2) = \mathbf{p}_2 \cos^2\frac{\theta}{2} + 2\mathbf{q}\mathbf{p}_2 \cos\frac{\theta}{2} - \mathbf{p}_2 \sin^2\frac{\theta}{2}$$

$$= -\mathbf{p}_1 \sin\theta + \mathbf{p}_2 \cos\theta$$

and

$$\rho_q(\mathbf{p}_3) = \rho_q(\mathbf{p}_1) \times \rho_q(\mathbf{p}_2)$$
$$= (\mathbf{p}_1 \times \mathbf{p}_2)\cos^2\theta - (\mathbf{p}_2 \times \mathbf{p}_1)\sin^2\theta$$
$$= \mathbf{p}_3.$$

It follows that the matrix representing the linear transformation ρ_q relative to the orthonormal basis $(\mathbf{p}_1, \mathbf{p}_2, \mathbf{p}_3)$ is

$$\begin{pmatrix} \cos\theta & -\sin\theta & 0 \\ \sin\theta & \cos\theta & 0 \\ 0 & 0 & 1 \end{pmatrix}.$$

The underlying mapping acts as a rotation by an angle θ counterclockwise with axis pointing in the direction of $\mathbf{q}$.

Compared to Euler angles and matrices, the quaternion representation of rotations has the advantage that given a rotation in quaternion notation it is easy to find the angle and axis of rotation.

Example. Let α and β be real numbers. Consider the two rotations $q_1 = \cos\frac{\alpha}{2} + \sin\frac{\alpha}{2}i$ and $q_2 = \cos\frac{\beta}{2} - \frac{1}{\sqrt{2}}\sin\frac{\beta}{2}(j - k)$. Show that the angle of rotation θ induced by the mapping $\rho_{(q_1 q_2)^{-1}}$ satisfies the relation

$$2\cos\theta = \cos\alpha\cos\beta + \cos\alpha + \cos\beta - 1.$$

Solution. To start with, the quaternion that describes the composite rotation operator $\rho_{(q_1 q_2)^{-1}}$ is

$$q_1 q_2 = \cos\frac{\alpha}{2}\cos\frac{\beta}{2} + \sin\frac{\alpha}{2}\cos\frac{\beta}{2}i - \frac{1}{\sqrt{2}}\left(\cos\frac{\alpha}{2} + \sin\frac{\alpha}{2}\right)\sin\frac{\beta}{2}j$$
$$+ \frac{1}{\sqrt{2}}\left(\cos\frac{\alpha}{2} - \sin\frac{\alpha}{2}\right)\sin\frac{\beta}{2}k.$$

Hence the axis of the composite rotation is defined by the vector

$$\left(\sin\frac{\alpha}{2}\cos\frac{\beta}{2}, -\frac{1}{\sqrt{2}}\left(\cos\frac{\alpha}{2} + \sin\frac{\alpha}{2}\right)\sin\frac{\beta}{2}, \frac{1}{\sqrt{2}}\left(\cos\frac{\alpha}{2} - \sin\frac{\alpha}{2}\right)\sin\frac{\beta}{2}\right),$$

and, moreover, the angle of rotation θ satisfies the conditions

$$\cos\frac{\theta}{2} = \cos\frac{\alpha}{2}\cos\frac{\beta}{2}, \quad \text{and} \quad \sin\frac{\theta}{2} = \sqrt{1 - \left(1 - \sin^2\frac{\alpha}{2}\right)\cos^2\frac{\alpha}{2}}.$$

By direct computation, one can show that

$$\cos \theta = 2\cos^2 \frac{\alpha}{2} \cos^2 \frac{\beta}{2} - 1 = \frac{\cos\alpha\cos\beta + \cos\alpha + \cos\beta - 1}{2}.$$

Now let p be an arbitrary quaternion and q_1, q_2 be two unit quaternions. As the next example shows, in the case when $\mathbf{q}_1$ and $\mathbf{q}_2$ are orthogonal, one can easily identify the combined rotation $(\rho_{q_1} \circ \rho_{q_2})(p)$ (about axes $\mathbf{q}_1$, $\mathbf{q}_2$ with respective angles θ_1, θ_2).

Example. With this insight, we shall now show that if $\mathbf{q}_1$ and $\mathbf{q}_2$ are orthogonal then $\rho_{q_1 q_2}(p)$ is the rotation through an angle θ with $\cos \frac{\theta}{2} = \cos \frac{\theta_1}{2} \cos \frac{\theta_2}{2}$, and parallel to the axis

$$\sin \frac{\theta_1}{2} \cos \frac{\theta_2}{2} i + \cos \frac{\theta_1}{2} \sin \frac{\theta_2}{2} j + \sin \frac{\theta_1}{2} \sin \frac{\theta_2}{2} k.$$

Solution. Without loss of generality, we can assume that $\mathbf{q}_1$ and $\mathbf{q}_2$ are parallel to the coordinate axes x and y, where (x, y, z, t) are the axes of $\mathbb{R}^4$. Then

$$q_1 = \cos \frac{\theta_1}{2} + \sin \frac{\theta_1}{2} i \quad \text{and} \quad q_2 = \cos \frac{\theta_2}{2} + \sin \frac{\theta_2}{2} j.$$

Hence the product $q_1 q_2$ is again a unit quaternion, and is equal to

$$\cos \frac{\theta_1}{2} \cos \frac{\theta_2}{2} + \sin \frac{\theta_1}{2} \cos \frac{\theta_2}{2} i + \cos \frac{\theta_1}{2} \sin \frac{\theta_2}{2} j + \sin \frac{\theta_1}{2} \sin \frac{\theta_2}{2} k.$$

Let's now consider an example in which we make use of the quaternion representation of rotations.

Example. Consider the rotation about the axis defined by $(0, 0, 1, 0)$ through the angle of $\frac{\pi}{2}$. Compute the effect of rotation on the basis vector $i = (0, 1, 0, 0)$.

Solution. The quaternion q that defines the rotation is here

$$q = \cos \frac{\pi}{4} + \sin \frac{\pi}{4} j = \frac{\sqrt{2}}{2} + \frac{\sqrt{2}}{2} j.$$

We now compute the effect of rotation on the basis vector $i = (0, 1, 0, 0)$. We get

$$q i q^{-1} = \left(\frac{\sqrt{2}}{2} + \frac{\sqrt{2}}{2} j \right) i \left(\frac{\sqrt{2}}{2} - \frac{\sqrt{2}}{2} j \right) = -k.$$

Exercise 2.68. *Consider the rotation q around the axis $\mathbf{p} = i + j + k$ with the angle of rotation $\theta = \frac{\pi}{2}$. Find q.*

Solution. $q = \frac{\sqrt{2}}{2} + \frac{\sqrt{6}}{6}(i + j + k)$.

Exercise 2.69. *Let $\theta \geq 0$. Prove that any unit quaternion $q = \cos\frac{\theta}{2} + \frac{\mathbf{p}}{|\mathbf{p}|}\sin\frac{\theta}{2}$ such that $\mathbf{p} = bi + cj + dk$, corresponds to the rotation $\rho_q(p)$ of matrix $I_3 + \frac{\sin\theta}{\theta}A + \frac{1-\cos\theta}{\theta^2}A$, where*

$$A = \begin{pmatrix} 0 & -d & c \\ d & 0 & -b \\ -c & b & 0 \end{pmatrix}.$$

Hint. In case $\theta = 0$, take the limit $\theta \to 0^+$.

If one looks back at the history of the development of the theory of three-dimensional rotations, the Euler-Rodrigues rotation formula clearly stands out. It provides an efficient tool for rotating a vector in space, given an axis and an angle of rotation.

2.5 Euler-Rodrigues Formula

Let q be a unit quaternion. Let the pure quaternion $\mathbf{u}$ be mapped by means of ρ_q, and let $\theta \in (0, \pi]$ be the required angle of rotation. The Euler-Rodrigues formula reads

$$\rho_q(\mathbf{u}) := (1 - \cos\theta)\left(\frac{\mathbf{p}}{|\mathbf{p}|} \cdot \mathbf{u}\right)\frac{\mathbf{p}}{|\mathbf{p}|} + \mathbf{u}\cos\theta + \left(\frac{\mathbf{p}}{|\mathbf{p}|} \times \mathbf{u}\right)\sin\theta. \qquad (2.4)$$

Write the unit quaternion $q = q_0 + \mathbf{q}$ as $q = \cos\frac{\theta}{2} + \frac{\mathbf{p}}{|\mathbf{p}|}\sin\frac{\theta}{2}$. Substituting $q_0 = \cos\frac{\theta}{2}$ and $\mathbf{q} = \frac{\mathbf{p}}{|\mathbf{p}|}\sin\frac{\theta}{2}$ in (2.1), a direct computation shows that

$$\rho_q(\mathbf{u}) = \left(\cos^2\frac{\theta}{2} - \frac{\mathbf{p}}{|\mathbf{p}|}\sin\frac{\theta}{2} \cdot \frac{\mathbf{p}}{|\mathbf{p}|}\sin\frac{\theta}{2}\right)\mathbf{u} + 2\left(\frac{\mathbf{p}}{|\mathbf{p}|}\sin\frac{\theta}{2} \cdot \mathbf{u}\right)\frac{\mathbf{p}}{|\mathbf{p}|}\sin\frac{\theta}{2}$$

$$+ 2\cos\frac{\theta}{2}\left(\frac{\mathbf{p}}{|\mathbf{p}|}\sin\frac{\theta}{2} \times \mathbf{u}\right)$$

$$= \left[\cos^2\frac{\theta}{2} - \sin^2\frac{\theta}{2}\left(\frac{\mathbf{p}}{|\mathbf{p}|} \cdot \frac{\mathbf{p}}{|\mathbf{p}|}\right)\right]\mathbf{u} + 2\sin^2\frac{\theta}{2}\left(\frac{\mathbf{p}}{|\mathbf{p}|} \cdot \mathbf{u}\right)\frac{\mathbf{p}}{|\mathbf{p}|}$$

$$+ 2\cos\frac{\theta}{2}\sin\frac{\theta}{2}\left(\frac{\mathbf{p}}{|\mathbf{p}|} \times \mathbf{u}\right).$$

Therefore

$$
\rho_q(\mathbf{u}) = \left(\cos^2\frac{\theta}{2} - \sin^2\frac{\theta}{2}\right)\mathbf{u} + 2\sin^2\frac{\theta}{2}\left(\frac{\mathbf{p}}{|\mathbf{p}|}\cdot\mathbf{u}\right)\frac{\mathbf{p}}{|\mathbf{p}|}
$$

$$
+ 2\cos\frac{\theta}{2}\sin\frac{\theta}{2}\left(\frac{\mathbf{p}}{|\mathbf{p}|}\times\mathbf{u}\right)
$$

$$
= (1-\cos\theta)\left(\frac{\mathbf{p}}{|\mathbf{p}|}\cdot\mathbf{u}\right)\frac{\mathbf{p}}{|\mathbf{p}|} + \mathbf{u}\cos\theta + \left(\frac{\mathbf{p}}{|\mathbf{p}|}\times\mathbf{u}\right)\sin\theta.
$$

Exercise 2.70. *Let q be a unit quaternion, $\mathbf{p}$ a pure quaternion with the angle of rotation π. Using the Euler-Rodrigues formula, prove that if $\rho_q(\mathbf{p}) = \mathbf{p}$, then $\mathbf{p}$ is orthogonal to* $\mathbf{q}$.

Exercise 2.71. *Let q be a unit quaternion and $\mathbf{p}$ be a pure quaternion such that $\mathbf{p} = \lambda\mathbf{q}$ for some $\lambda \in \mathbb{R}$. Using the Euler-Rodrigues formula, prove that $\rho_q(\mathbf{p}) = \lambda\,\mathbf{q}$.*

2.6 Applications to Plane Geometry

Let q be a unit quaternion. Given a pure quaternion $\mathbf{u}$ we can actually decompose it as $\mathbf{u} = \mathbf{a} + \mathbf{n}$, where $\mathbf{a}$ is the component along $\mathbf{q}$ and $\mathbf{n}$ is the component normal to $\mathbf{q}$. We will show that under the mapping ρ_q, $\mathbf{n}$ is rotated about q through an angle θ. Since the operator ρ_q is linear, the image $q\mathbf{u}\overline{q}$ can be indeed interpreted as the result of rotating $\mathbf{u}$ about $\mathbf{q}$ through an angle θ. Since $\mathbf{a}$ is invariant under ρ_q, we focus on how ρ_q acts on the orthogonal component $\mathbf{n}$. Simple calculations give

$$
\rho_q(\mathbf{n}) = q\mathbf{n}\overline{q}
$$

$$
= (q_0 + \mathbf{q})[\mathbf{q}\cdot\mathbf{n} + (\mathbf{n}q_0 - \mathbf{n}\times\mathbf{q})]
$$

$$
= q_0(\mathbf{q}\cdot\mathbf{n}) + q_0(\mathbf{n}q_0 - \mathbf{n}\times\mathbf{q}) + \mathbf{q}(\mathbf{q}\cdot\mathbf{n}) + \mathbf{q}(\mathbf{n}q_0 - \mathbf{n}\times\mathbf{q}).
$$

From (1.2) it follows that

$$
\rho_q(\mathbf{n}) = q_0(\mathbf{q}\cdot\mathbf{n}) + q_0^2\mathbf{n} - q_0(\mathbf{n}\times\mathbf{q}) + \mathbf{q}(\mathbf{q}\cdot\mathbf{n}) - \mathbf{q}\cdot(\mathbf{n}q_0 - \mathbf{n}\times\mathbf{q})
$$

$$
+ \mathbf{q}\times(\mathbf{n}q_0 - \mathbf{n}\times\mathbf{q})
$$

$$
= q_0(\mathbf{q}\cdot\mathbf{n}) + q_0^2\mathbf{n} - q_0(\mathbf{n}\times\mathbf{q}) + \mathbf{q}(\mathbf{q}\cdot\mathbf{n}) - (\mathbf{q}\cdot\mathbf{n})q_0 + \mathbf{q}\cdot(\mathbf{n}\times\mathbf{q})
$$

$$
+ \mathbf{q}\times(\mathbf{n}q_0) - \mathbf{q}\times(\mathbf{n}\times\mathbf{q})
$$

$$
= q_0^2\mathbf{n} - q_0(\mathbf{n}\times\mathbf{q}) + (\mathbf{q}\cdot\mathbf{n})\mathbf{q} + \mathbf{q}\times(\mathbf{n}q_0) - \mathbf{q}\times(\mathbf{n}\times\mathbf{q})
$$

$$
= q_0^2\mathbf{n} + (\mathbf{q}\cdot\mathbf{n})\mathbf{q} - \mathbf{q}\times(\mathbf{n}\times\mathbf{q}) - 2q_0(\mathbf{n}\times\mathbf{q}).
$$

Applying formula (1.3) to the previous expression we obtain

$$\rho_q(\mathbf{n}) = q_0^2 \mathbf{n} + (\mathbf{q} \cdot \mathbf{n})\mathbf{q} - (\mathbf{q} \cdot \mathbf{q})\mathbf{n} + (\mathbf{q} \cdot \mathbf{n})\mathbf{q} + 2q_0(\mathbf{q} \times \mathbf{n})$$

$$= (q_0^2 - \mathbf{q} \cdot \mathbf{q})\mathbf{n} + 2(\mathbf{q} \cdot \mathbf{n})\mathbf{q} + 2q_0(\mathbf{q} \times \mathbf{n})$$

$$= (q_0^2 - |\mathbf{q}|^2)\mathbf{n} + 2q_0(\mathbf{q} \times \mathbf{n})$$

$$= (q_0^2 - |\mathbf{q}|^2)\mathbf{n} + 2q_0|\mathbf{q}| \left(\frac{\mathbf{q}}{|\mathbf{q}|} \times \mathbf{n} \right).$$

Denote $\mathbf{n}_\perp = \frac{\mathbf{q}}{|\mathbf{q}|} \times \mathbf{n}$. With these calculations at hand we set

$$\rho_q(\mathbf{n}) = (q_0^2 - |\mathbf{q}|^2)\mathbf{n} + 2q_0|\mathbf{q}|\mathbf{n}_\perp.$$

Note that $\mathbf{n}_\perp$ and $\mathbf{n}$ have the same length:

$$|\mathbf{n}_\perp| = \left| \mathbf{n} \times \frac{\mathbf{q}}{|\mathbf{q}|} \right| = |\mathbf{n}| \left| \frac{\mathbf{q}}{|\mathbf{q}|} \right| \sin \frac{\pi}{2} = |\mathbf{n}|.$$

Ultimately, we arrive at

$$\rho_q(\mathbf{n}) = \left(\cos^2 \frac{\theta}{2} - \sin^2 \frac{\theta}{2} \right) \mathbf{n} + \left(2 \cos \frac{\theta}{2} \sin \frac{\theta}{2} \right) \mathbf{n}_\perp$$

$$= \cos \theta \, \mathbf{n} + \sin \theta \, \mathbf{n}_\perp.$$

Namely, the resulting vector is a rotation of $\mathbf{n}$ through an angle θ in the plane defined by $\mathbf{n}$ and $\mathbf{n}_\perp$. This vector is clearly orthogonal to the rotation axis. The vector resulting from rotating a vector $\mathbf{u}$ about the axis $\frac{\mathbf{p}}{|\mathbf{p}|}$ through θ is thus given by (2.4).

Example. Let $q = \frac{1}{\sqrt{3}}(i - j + k)$ and $\mathbf{n} = j + k$. Find $\mathbf{n}_\perp$ and $\rho_q(\mathbf{n})$.

Solution. Straightforward computations show that

$$\mathbf{n}_\perp = \frac{\mathbf{q}}{|\mathbf{q}|} \times \mathbf{n} = -\frac{2}{\sqrt{3}}i - \frac{1}{\sqrt{3}}j + \frac{1}{\sqrt{3}}k, \qquad \rho_q(\mathbf{n}) = -|\mathbf{q}|^2\mathbf{n} = -j - k.$$

Exercise 2.72. *Let* $q = \frac{1}{2}(1 - i + j + k)$ *and* $\mathbf{n} = i + k$. *Represent* $\rho_q(\mathbf{n})$ *in trigonometric form.*

Solution. $\rho_q(\mathbf{n}) = \sqrt{2}\left(\cos \frac{\pi}{2} - \frac{j-k}{\sqrt{2}} \sin \frac{\pi}{2} \right).$

Exercise 2.73. *Let* $\rho_q(\mathbf{n}) = i - j + k$ *and* $q = \frac{1}{\sqrt{3}}(1 + i + j)$. *Find* $\mathbf{n}$.

Solution. $\mathbf{n} = -i + j + k$.

Exercise 2.74. *Let $\rho_q(\mathbf{n}) = 2i + j + k$ and $\mathbf{n} = i - j$. Find q.*

Solution. No solution.

Exercise 2.75. *Let $p_l \in \mathbb{H}$ ($l = 1, \ldots, n$) be such that $p_l \cdot p_m = 0$ when $l \neq m$ for $l, m \in \{1, 2, \ldots, n\}$. Prove that $|p_1 + \cdots + p_n|^2 = |p_1|^2 + \cdots + |p_n|^2$.*

2.7   Advanced Practical Exercises

1. Find $\rho_q(p_1) \cdot \rho_q(p_2)$ if (a) $p_1 = 1 - i - j - k$, $p_2 = i + j + k$, $q = 1 + i + j + k$; (b) $p_1 = 2 - 3i - 4j - 2k$, $p_2 = 1 + i + j$, $q = 2 + 7i + j + 5k$; (c) $p_1 = 7 - i + k$, $p_2 = 49 - 48j$, $q = 101 + 102i + 103j + 104k$.
2. Find $|\rho_q(p)|$ if (a) $p = 1 + 2i - j - k$, $q = 1 + i + j$; (b) $p = 1 - 3i + 2j - k$, $q = 1 - i - k$; (c) $p = 1 - 4i + 5j + 2k$, $q = 2 - 101i + 203j + 4k$.
3. Let $p = 1 - i - j - k$, $q_1 = 1 + i$ and $q_2 = 1 - j$. Find $\rho_{q_1} \rho_{q_2}(p)$.
4. Let $q = 1 + i + j + k$. Find $p \in \mathbb{H}$ such that $\rho_q(p) = j$.
5. Rotate the vector $\mathbf{x} = (x_1 \ \ x_2 \ \ x_3)^T$ by the angle θ about the unit vector $\mathbf{n} = (n_1 \ \ n_2 \ \ n_3)^T$ and denote the resulting vector by $\mathbf{x}'$.
   (a) Prove that $\mathbf{x}' = (1 - \cos\theta)(\mathbf{x} \cdot \mathbf{n})\mathbf{n} + \cos\theta \mathbf{x} + (\mathbf{n} \times \mathbf{x})\sin\theta$;
   (b) Find an 3×3 matrix $R(\mathbf{n}, \theta)$ so that $\mathbf{x}' = R(\mathbf{n}, \theta)\mathbf{x}$;
   (c) Find an 3×3 matrix N so that $N\mathbf{x} = \mathbf{n} \times \mathbf{x}$;
   (d) Find the matrix N^2;
   (e) Prove that $N^3 = -N$;
   (f) Prove that $R(\mathbf{n}, \theta) = I + \sin\theta N + (1 - \cos\theta)N^2$;
   (g) Prove that $R(\mathbf{n}, \theta) = \left(R\left(\mathbf{n}, \frac{\theta}{k}\right)\right)^k \quad \forall k \in \mathbb{N}$;
   (h) Prove that $R(\mathbf{n}, \theta) = e^{\theta N}$;
   (i) Prove that $R(\mathbf{n}, \theta) = I + \sin\theta N + (1 - \cos\theta)(\mathbf{n}\mathbf{n} - I)$.
6. Let $\mathbf{u}$ and $\mathbf{v}$ be pure quaternions. Prove the following assertions: (a) $\mathbf{u}\mathbf{v} - \mathbf{v}\mathbf{u} = 2(\mathbf{u} \times \mathbf{v})$;   (b) $\mathbf{u}\mathbf{v}\mathbf{u} = \mathbf{v}(\mathbf{u} \cdot \mathbf{u}) - 2\mathbf{u}(\mathbf{u} \cdot \mathbf{v})$.
7. Let $\mathbf{u}$ be a pure quaternion such that $|\mathbf{u}| = 1$. Let $\mathbf{v}$ be an ordinary vector in $\mathbb{R}^3$, and $\mathbf{v}'$ the vector obtained by rotating $\mathbf{v}$ through an angle θ around the axis $\mathbf{u}$. Prove that $\mathbf{v}' = q\mathbf{v}\overline{q}$, where $q = \cos\frac{\theta}{2} + \mathbf{u}\sin\frac{\theta}{2}$.
8. Consider the rotation f around the axis $\mathbf{v} = i + j + k$ with rotation angle $\theta = \frac{2\pi}{3}$. Find $f(x)$, where (a) $x = i$;   (b) $x = j$;   (c) $x = k$;   (d) $x = i + j$; (e) $x = i + k$;   (f) $x = j + k$;   (g) $x = i + j + k$;   (h) $x = i - j - k$;   (i) $x = 2i - 3j + k$;   (j) $x = -1 + i + j - 4k$;   (k) $x = ai + bj + ck$ ($a, b, c \in \mathbb{R}$);   (l) $x = a_0 + a + bi + cj$ ($a_0, a, b, c \in \mathbb{R}$).
9. Consider the rotation f around the axis $\mathbf{u} = -i + \sqrt{2}j + k$ with rotation angle $\theta = \frac{\pi}{2}$. Find $f(x)$, where (a) $x = i$;   (b) $x = j$;   (c) $x = k$; (d) $x = i - j + k$;   (e) $x = 1 + i + j + k$;   (f) $x = ai + bj + ck$ ($a, b, c \in \mathbb{R}$); (g) $x = a_0 + ai + bj + ck$ ($a_0, a, b, c \in \mathbb{R}$).

10. Let $p = 1 + i - j - k$, $q = 3 + 2i + j + 4k$ and $r = 2 + i + j + k$. Consider the rotation f around the axis $\mathbf{u} = i + j + k$ with rotation angle $\theta = \frac{2\pi}{3}$. Find:
    (a) $f(2p - pqr^2)$;   (b) $f(p + q^3 + pqr^2)$.

11. Let $p = a_1 + b_1 i + c_1 j + d_1 k$ and $q = a_2 + b_2 i + c_2 j + d_2 k$ with $a_i, b_i, c_i, d_i \in \mathbb{R}$ $(i = 1, 2)$. Consider the rotation f around the axis $\mathbf{u} = i + j + k$ with rotation angle $\theta = \frac{2\pi}{3}$. Find (a) $f((p, q))$;   (b) $f([p, q])$;   (c) $f(p \times q)$;
    (d) $f(\mathrm{sgn}(p))$;   (e) $f(p_+)$;   (f) $f(p_-)$.

12. Let $p = i + j$, $q = 1 - i - k$ and $r = i + j + k$. Consider the rotation f around the axis $\mathbf{u} = i + j + k$ with rotation angle $\theta = \frac{2\pi}{3}$. Find $f(p - [p, 2p - r] + q \times (p + r))$.

13. Consider the rotation f around the axis $\mathbf{u} = i + j + k$ with rotation angle $\theta = \frac{2\pi}{3}$. Find $f(p)$, where (a) $p = j^k(1 + i - j) + 2 - i - 3j + k + [i + j, i - j]$;
    (b) $p = \sin(i) + \cos(j) + \tan(k)$.

14. Consider the rotation f around the axis $\mathbf{v} = v_1 i + v_2 j + v_3 k$, $v_1, v_2, v_3 \in \mathbb{R}$, with rotation angle θ. Prove that for every $x \in \mathbb{H}$, it holds that $f(x) = f(-x)$.

15. Show that the quaternion product is noncommutative using rotations.

16. Let $q = q_0 + \mathbf{q} \in \mathbb{H}$ be a unit quaternion, $x = x_0 + \mathbf{x} \in \mathbb{H}$. We define the quaternion $x' = qx\overline{q}$ with $x' = x_0' + \mathbf{x}'$, where $\mathbf{x}' = x_1' i + x_2' j + x_3' k$ $(x_0', x_1', x_2', x_3' \in \mathbb{R})$.
    (a) Prove that $x_0' = x_0$;
    (b) Prove that $\mathbf{x}' \cdot \mathbf{x}' = \mathbf{x} \cdot \mathbf{x}$;
    (c) Prove that $\mathbf{x}' = (q_0^2 - \mathbf{q} \cdot \mathbf{q})\mathbf{x} + 2q_0\mathbf{q} \times \mathbf{x} + 2\mathbf{q}(\mathbf{q} \cdot \mathbf{x})$;
    (d) Find an 3×3 matrix $R(q)$ such that $\mathbf{x}' = R(q)\mathbf{x}$;
    (e) Prove that the correspondence $q \to R(q)$ is a group homomorphism;
    (f) Prove that the correspondence $q \to R(q)$ is not an isomorphism;
    (g) Prove that $R(q)$ is an orthogonal matrix;
    (h) Let $R = (r_{ij})$. Prove that

$$q_0^2 = \frac{1}{4}(1 + r_{11} + r_{22} + r_{33}), \qquad q_1^2 = \frac{1}{4}(1 + r_{11} - r_{22} - r_{33}),$$

$$q_2^2 = \frac{1}{4}(1 - r_{11} + r_{22} - r_{33}), \qquad q_3^2 = \frac{1}{4}(1 - r_{11} - r_{22} + r_{33}),$$

$$q_0 q_1 = \frac{1}{4}(r_{32} - r_{23}), \qquad q_0 q_2 = \frac{1}{4}(r_{13} - r_{31}),$$

$$q_0 q_3 = \frac{1}{4}(r_{21} - r_{12}), \qquad q_1 q_2 = \frac{1}{4}(r_{12} + r_{21}),$$

$$q_1 q_3 = \frac{1}{4}(r_{13} + r_{31}), \qquad q_2 q_3 = \frac{1}{4}(r_{23} + r_{32}).$$

17. Prove that in any right-angled triangle the square on the hypothenuse is equal to the sum of the square on the sides.

18. Prove that in any right-angled triangle the median to the hypothenuse is one-half the hypothenuse.

19. Let $ABCD$ be a parallelogram. If its diagonals are at right angles to each other, it is a rhombus.
20. The figure formed by joining the middle points of the sides of a square is itself a square.
21. The sines of the angles in any plane triangle are proportional to the opposite sides.
22. In a right-angled triangle find the sine and cosine of the acute angle.
23. Find the sine and cosine of the sum of two angles.
24. Find the angle between the diagonals of a parallelogram.
25. Prove that the sum of the squares of the diagonals of a parallelogram equals the sum of the squares of the sides.
26. Prove that in any quadrilateral, if the lines joining the middle points of opposite sides are at right angles, then the diagonals are equal.
27. Inscribe a circle in a given triangle. Prove that the sum of the angles of the triangle is $180°$.

Hint. Let α, β, γ be the angles of the triangle. Note that the corresponding central angles are $180° - \alpha$, $180° - \beta$ and $180° - \gamma$, and their sum is $360°$.

Quaternion Sequences

3

In the present chapter we use the properties of quaternions described in a previous chapter to explore the key notion of a quaternion sequence. Then we will use this analogue in a formula called summation by parts, which is an analogue of integration by parts for sums. Summation by parts is not only a useful auxiliary tool, but even indispensable in many applications, including finding sums of powers of integers and deriving some famous convergence tests for series: the Dirichlet and Abel tests. Besides the new results that we will derive from it, much of the theory of quaternion sequences is analogous to that encountered in real and complex calculus. For this reason, the prerequisites are rather modest. A strong motivation for taking up this study is that we need to understand the behavior of quaternion sequences in order to construct formal quaternion power series and infinite products in Chap. 4.

3.1 Quaternion Sequence

A quaternion sequence is a collection of real quaternions $p_1, p_2, p_3, \ldots$ "labelled" by nonnegative integers. We shall denote such a sequence by $\{p_n\}_{n=1}^{\infty}$, where $n = 1, 2, 3, \ldots$ and p_n are the elements of the sequence. As an alternative simpler notation from now on we write $\{p_n\}$ for $\{p_n\}_{n=1}^{\infty}$. Traditionally, one assumes that a quaternion sequence has infinitely many terms; a finite collection is sometimes called a finite quaternion sequence. There is no requirement for the terms of a quaternion sequence to be distinct, some of them may coincide. If all terms p_n coincide with the same quaternion number, the sequence is called constant. To give an example, we take $p_n = n - j^n + k^n$ so that the quaternion sequence $\{n - j^n + k^n\}$ is

$$\underbrace{1 - j + k,}_{n=1} \quad \underbrace{2}_{n=2}, \quad \underbrace{3 + j - k,}_{n=3} \quad \underbrace{4}_{n=4}, \quad \underbrace{5 - j + k,}_{n=5} \quad \ldots .$$

Let $\{p_n\}$ be a quaternion sequence. We make the following definitions:

J.P. Morais et al., *Real Quaternionic Calculus Handbook*,
DOI 10.1007/978-3-0348-0622-0_3, © Springer Basel 2014

(i) We say that $\{p_n\}$ is nonzero if $p_n \neq 0_{\mathbb{H}}$ for every $n \in \mathbb{N}$;
(ii) We say that two quaternion sequences $\{p_n\}$ and $\{q_n\}$ are equal, and write $\{p_n\} = \{q_n\}$ if $p_n = q_n$ for all $n \in \mathbb{N}$.

3.2   Scalar and Vector Parts of a Quaternion Sequence

Let $\{p_n\} := \{a_n\} + \{b_n\}i + \{c_n\}j + \{d_n\}k$ be a quaternion sequence. The sequence $\{a_n\}$ denotes the scalar part of $\{p_n\}$ and $\{\mathbf{p}_n\} := \{b_n\}i + \{c_n\}j + \{d_n\}k$ the vector part of $\{p_n\}$. Here $\{a_n\}, \{b_n\}, \{c_n\}$ and $\{d_n\}$ are real-valued sequences defined on a set of positive integers. The notations $\{a_n\}$ and $\{\mathrm{Sc}(p_n)\}$ (resp. $\{\mathbf{p}_n\}$ and $\{\mathrm{Vec}(p_n)\}$) are interchangeable.

3.3   Symmetric, Conjugate, Modulus and Inverse
of a Quaternion Sequence

Given a quaternion sequence $\{p_n\}$, we define:
  (i) Symmetric
    $-\{p_n\} = \{-p_n\}$;
 (ii) Conjugate
    $\overline{\{p_n\}} = \{\overline{p_n}\}$;
(iii) Inverse
    $\{p_n\}^{-1} = \{p_n^{-1}\}$, provided that $p_n \neq 0_{\mathbb{H}}$ for all n;
(iv) Modulus
    $|\{p_n\}| = \{|p_n|\}$.
On occasion the modulus of a quaternion sequence $\{p_n\} = \{a_n\} + \{b_n\}i + \{c_n\}j + \{d_n\}k$ is defined by

$$\sqrt{\{p_n\}\overline{\{p_n\}}} = \sqrt{\overline{\{p_n\}}\{p_n\}} = \sqrt{\{(a_n)^2\} + \{(b_n)^2\} + \{(c_n)^2\} + \{(d_n)^2\}}.$$

Exercise 3.76. *Show that the following statements are true:* (a) *if $\{p_n\}$ is nonzero then $\{-p_n\}$ is nonzero;*   (b) *if $\{p_n\}$ is nonzero then $\{p_n\}^{-1}$ is nonzero;*   (c) *if $\{p_n\}$ is nonzero then $(\{p_n\}^{-1})^{-1} = \{p_n\}$.*

Exercise 3.77. *Prove that* (a) *if $\lambda \in \mathbb{H} \setminus \{0_{\mathbb{H}}\}$ and $\{p_n\}$ is nonzero then $\{\lambda p_n\}$ is nonzero;*   (b) *if $\lambda \in \mathbb{H} \setminus \{0_{\mathbb{H}}\}$ and $\{p_n\}$ is nonzero then $\{\lambda p_n\} = \lambda\{p_n\}$;* (c) *if $\lambda \in \mathbb{H} \setminus \{0_{\mathbb{H}}\}$ and $\{p_n\}$ is nonzero then $\{\lambda p_n\}^{-1} = \{p_n\}^{-1}\lambda^{-1}$.*

3.4   Arithmetic Operations

As in the case with real and complex sequences, we have the standard operations on quaternion sequences. In particular, given two quaternion sequences $\{p_n\}$ and $\{q_n\}$, we define:

(i) Addition (subtraction)
$$\{p_n\} \pm \{q_n\} = \{p_n \pm q_n\};$$
(ii) Multiplication
$$\{p_n\}\{q_n\} = \{p_n q_n\};$$
(iii) Quotient
$$\frac{\{p_n\}}{\{q_n\}} = \left\{\frac{p_n}{q_n}\right\} = \{p_n q_n^{-1}\}, \text{ provided that } q_n \neq 0_{\mathbb{H}} \text{ for all } n.$$

We shall always write $\frac{\{p_n\}}{\{q_n\}}$ to mean $\{p_n\}\{q_n^{-1}\}$ in the sequel. Note that this is generally different from $\{q_n^{-1}\}\{p_n\}$, since quaternion multiplication is not commutative.

The familiar associativity, and distributivity laws of multiplication over addition (subtraction) hold for quaternion sequences:

(iv) Commutativity law of addition (subtraction)
$$\{p_n\} \pm \{q_n\} = \{q_n\} \pm \{p_n\} \text{ for all } \{p_n\}, \{q_n\} \subset \mathbb{H};$$
(v) Associativity law of addition (subtraction)
$$\{p_n\} \pm (\{q_n\} \pm \{r_n\}) = (\{p_n\} \pm \{q_n\}) \pm \{r_n\} \text{ for all } \{p_n\}, \{q_n\}, \{r_n\} \subset \mathbb{H};$$
(vi) Distributivity law of multiplication over addition (subtraction)
$$\{p_n\}(\{q_n\} \pm \{r_n\}) = \{p_n\}\{q_n\} \pm \{p_n\}\{r_n\} \text{ and}$$
$$(\{q_n\} \pm \{r_n\})\{p_n\} = \{q_n\}\{p_n\} \pm \{r_n\}\{p_n\} \text{ for all } \{p_n\}, \{q_n\}, \{r_n\} \subset \mathbb{H};$$
(vii) Associativity law of multiplication
$$(\{p_n\}\{q_n\})\{r_n\} = \{p_n\}(\{q_n\}\{r_n\}) \text{ for all } \{p_n\}, \{q_n\}, \{r_n\} \subset \mathbb{H}.$$
The verification of these properties is left to the reader.

Exercise 3.78. *Show that each of the following statements is true:* (a) $\{p_n\}$ *and* $\{q_n\}$ *are nonzero if and only if* $\{p_n q_n\}$ *is nonzero;* (b) *if* $\{p_n\}$ *and* $\{q_n\}$ *are nonzero, then* $\{p_n q_n\}^{-1} = \{q_n\}^{-1}\{p_n\}^{-1};$ (c) *if* $\{q_n\}$ *is nonzero, then* $\frac{\{p_n\}}{\{q_n\}}\{q_n\} = \{p_n\}$ *for every quaternion sequence* $\{p_n\};$ (d) *if* $\{p_n\}$ *and* $\{q_n\}$ *are nonzero, then* $\frac{\{p_n\}}{\{q_n\}}$ *is nonzero;* (e) *if* $\{p_n\}$ *and* $\{q_n\}$ *are nonzero, then* $\left(\frac{\{p_n\}}{\{q_n\}}\right)^{-1} = \frac{\{q_n\}}{\{p_n\}};$ (f) *if* $\{p_n\}$ *is nonzero, then* $|\{p_n\}|$ *is nonzero;* (g) *if* $\{p_n\}$ *is nonzero, then* $(\{|p_n|\})^{-1} = |\{p_n^{-1}\}|.$

Exercise 3.79. *Let* $p_n = 1^n + i^n - j$ *and* $q_n = 2^n + i^n - k.$ *Find* (a) $\{p_n + q_n\};$ (b) $\{p_n - q_n\};$ (c) $\{p_n q_n\};$ (d) $\{p_n q_n^{-1}\};$ (e) $\overline{\{p_n\}};$ (f) $\overline{\{q_n\}};$ (g) $|\{p_n\}|;$ (h) $|\{q_n\}|.$

Solution. (a) $1 + 2^n + 2i^n - j - k;$ (b) $1 - 2^n - j - k;$ (c) $2^n + i^n - k + (2i)^n + (-1)^n - ki^n - j2^n - ji^n + i;$ (d) $\frac{2^n - i^n + k + 2^n i^n + (-1)^{n+1} + ki^n - j2^n + ji^n - i}{2^{2n} + 2};$ (e) $1^n - 2^n + j;$ (f) $2^n - i^n + k;$ (g) $\sqrt{3};$ (h) $\sqrt{2^{2n} + 2}.$

Exercise 3.80. *Let* $\{p_n\}$ *and* $\{q_n\}$ *be quaternion sequences. Show that:* (a) $|\{\lambda p_n\}| = |\lambda||\{p_n\}|$ *for every* $\lambda \in \mathbb{H} \setminus \{0_{\mathbb{H}}\};$ (b) $|\{p_n q_n\}| = |\{p_n\}||\{q_n\}|;$ (c) *If* $\{p_n\}$ *is nonzero, then* $\left|\frac{\{q_n\}}{\{p_n\}}\right| = \frac{|\{q_n\}|}{|\{p_n\}|}.$

Exercise 3.81. *Let* $p_n = ni - \frac{1}{n}k$ *and* $q_n = \frac{1}{n}i - nk$. *Find* (a) $|\{jp_n\}|$; (b) $|\{p_n q_n\}|$; (c) $\left|\frac{\{q_n\}}{\{p_n\}}\right|$.

Solution. (a) $\left\{\sqrt{n^2 + \frac{1}{n^2}}\right\}$; (b) $\left\{\sqrt{2 + n^4 + \frac{1}{n^4}}\right\}$; (c) $\left\{\frac{n^2}{1+n^4}\sqrt{2 + n^4 + \frac{1}{n^4}}\right\}$.

We proceed to study the convergence and divergence of quaternion sequences. Here we will recall some fundamental definitions and notations which will be needed throughout the text, probably familiar to most readers in one form or other.

3.5 Convergence of a Quaternion Sequence

The notion of convergence of a sequence is the same as that of a limit. We say that the sequence of quaternion numbers $\{p_n\}$ converges to a limit $p \in \mathbb{H}$ if $|p_n - p| \to 0$ as $n \to \infty$. We will use the traditional notation: $\lim_{n\to\infty} p_n = p$, or $p_n \to p$ as $n \to \infty$. In other words, $\{p_n\}$ converges to the real quaternion p if for each positive quantity ϵ, no matter how small, a natural number N can be found such that $|p_n - p| < \epsilon$ whenever $n > N$. This definition makes decisive use of the absolute value. Since the notion of absolute value has a meaning for quaternions as well as for real and complex numbers, we can use the same definition regardless of whether the variable p and the functions studied are real, complex or quaternion.

3.6 Divergence of a Quaternion Sequence

The sequence $\{p_n\} = \{a_n\} + \{b_n\}i + \{c_n\}j + \{d_n\}k$ of real quaternions is called divergent to infinity if for every positive constant M_1 (no matter how large), there exists a natural number $N = N(M_1)$ such that $a_n, b_n, c_n, d_n > M_1$ for every $n > N$. For this one uses the notation: $\lim_{n\to\infty} p_n = \infty$, or $p_n \to \infty$ as $n \to \infty$. The sequence $\{p_n\}$ is called divergent to negative infinity, and we write $\lim_{n\to\infty} p_n = -\infty$, or $p_n \to -\infty$ as $n \to \infty$, if for every constant $M_2 \in \mathbb{R}$, there exists $N = N(M_2)$ such that $a_n, b_n, c_n, d_n < M_2$ for every $n > N$. For example, the quaternion sequence $\{-n - n^2 i - n^3 j - nk\}$ is divergent to negative infinity.

The natural number N above certainly depends on the positive number ϵ. If ϵ_1 is a positive number smaller than ϵ, then the corresponding N_1 shall be larger than N. Hence we will often write N to mean the dependence $N(\epsilon)$. It should also be stressed that N is not unique. If a natural number N has the property needed in the definition of convergence, then any larger natural number M will also have that property. In practice, if we wish to attack a limit problem we will eventually have to show that the quantity $|p_n - p|$ is less than ϵ for some $p \in \mathbb{H}$ when $n \geq N$. To find out which N to use is always the challenging part. Commonly, some algebraic manipulations can be performed on the expression $|p_n - p|$ that can help us find out exactly how to choose N. As we will see later, there are ways to tell in advance that a quaternion sequence converges without knowing the value of the limit.

Let us list now some useful convergence properties. Let $\{p_n\}$ and $\{q_n\}$ be quaternion sequences such that $\lim_{n\to\infty} p_n = p$ and $\lim_{n\to\infty} q_n = q$, for $p, q \in \mathbb{H}$. Then

(i) $\lim_{n\to\infty} \overline{p_n} = \overline{p}$;

(ii) $\lim_{n\to\infty} |p_n| = |p|$;

(iii) $\lim_{n\to\infty} (p_n \pm q_n) = p \pm q$;

(iv) $\lim_{n\to\infty} p_n q_n = pq$;

(v) $\lim_{n\to\infty} p_n^{-1} = p^{-1}$ whenever $p \neq 0_\mathbb{H}$ and $p_n \neq 0_\mathbb{H}$ for all n.

The verification of these statements is not difficult and is left to the reader. Properties (iii) and (iv) can be extended to cope with any finite number of convergent quaternion sequences. For example, for n sequences: $\lim_{n\to\infty}(p_n \pm q_n \pm \cdots \pm r_n) = \lim_{n\to\infty} p_n \pm \lim_{n\to\infty} q_n \pm \cdots \pm \lim_{n\to\infty} r_n$, and $\lim_{n\to\infty}(p_n q_n \cdots r_n) = \lim_{n\to\infty} p_n \lim_{n\to\infty} q_n \cdots \lim_{n\to\infty} r_n$. Properties (iv) and (v) imply that $\lim_{n\to\infty} \frac{p_n}{q_n} = pq^{-1}$ whenever $q \neq 0_\mathbb{H}$ and $q_n \neq 0_\mathbb{H}$ for all n. Furthermore, if p is a quaternion number such that $|p| < 1$, then the quaternion sequence $\{(p_n)^n\}$ has limit $0_\mathbb{H}$. If $p \neq 0_\mathbb{H}$, then the sequence $\{|p_n|^{1/n}\}$ has limit 1.

Exercise 3.82. *Determine whether the following quaternion sequence converges or diverges:* (a) $\frac{n}{\sqrt[n]{n!}}(1 - i) + \frac{n}{n+1}(j - k)$; (b) $e^{1/n} - \left(1 + \frac{1}{n}\right)^n i + \frac{n}{\sqrt[n]{n!}}k$; (c) $\frac{n+(j+k)^n}{\sqrt{n}}$; (d) $\tan^{-1}(-n) - \left(1 - \frac{1}{n}\right)^n (i + j + k)^n$; (e) $i - j - n\log\left(\frac{2n+1}{2n-1}\right)k$; (f) $e^{-na}n^b \frac{i-j}{3+k}$ $(a, b > 0)$; (g) $n^{-a}\log(n)\frac{2+i-j+k}{2-i+j-k}$ $(a > 0)$; (h) $\frac{(n-nj+k)^2}{n^2} + \frac{\sqrt{n}}{2^{2n}}\frac{(2n)!}{(n!)^2}(i - 3j)$.

Solution. (a) Convergent; (b) Convergent; (c) Divergent; (d) Divergent; (e) Convergent; (f) Convergent; (g) Convergent; (h) Convergent.

Exercise 3.83. *Let $p_n, q_n \subset \mathbb{H}$ such that $\lim_{n\to\infty} p_n = p$ and $\lim_{n\to\infty} q_n = q$, $p, q \in \mathbb{H}$. Let $a, b \in \mathbb{H}$. Prove that $\lim_{n\to\infty}(ap_n \pm bq_n) = ap \pm bq$.*

Exercise 3.84. *Let $\{p_n\}$ be a quaternion sequence. Prove that $\lim_{n\to\infty} p_n = 0_\mathbb{H}$ if and only if $\lim_{n\to\infty} |p_n| = 0$.*

Exercise 3.85. *Let $\{p_n\}$ be a convergent quaternion sequence. Discuss the convergence of the quaternion sequence $n\{p_n\}^n$.*

Solution. The sequence is convergent if $|p_n| < \frac{1}{n^{\frac{1}{n}}}$.

Exercise 3.86. *Let $\{p_n\}$ be a quaternion sequence.* (a) *Suppose that, for each n, $|p_n| < 1/n$. Prove that $\lim_{n\to\infty} p_n = 0_\mathbb{H}$;* (b) *suppose $\{q_n\}$ is a quaternion* *sequence that converges to $0_{\mathbb{H}}$, and suppose that, for each n, $|p_n| < |q_n|$. Prove that* $\lim_{n\to\infty} p_n = 0_{\mathbb{H}}$.

We shall now state a necessary and sufficient condition for the existence of a limit for a sequence of quaternion numbers.

3.7   Criterion for Convergence

A quaternion sequence $\{p_n\}$ converges to a quaternion number $p = a + bi + cj + dk$ if and only if $\mathrm{Sc}(p_n)$ converges to $p_0 = a$ and $\mathrm{Vec}(p_n)$ converges to $\mathbf{p} = bi + cj + dk$.

For example, consider the sequence $\{\frac{n+i-j+nk}{n+1}\}$. We have

$$\mathrm{Sc}(p_n) = \frac{n}{n+1} \to 1$$

and

$$\mathrm{Vec}(p_n) = \frac{i - j + nk}{n+1} \to k$$

as $n \to \infty$. We conclude that the given sequence converges to $a + bi + cj + dk = 1 + k$.

Exercise 3.87. *Check whether the following quaternion sequence $\{p_n\}$ converges to a quaternion number p by computing $\lim_{n\to\infty} \mathrm{Sc}(p_n)$ and $\lim_{n\to\infty} \mathrm{Vec}(p_n)$.*
(a) $\{\frac{n^{-2}-n^3 k}{n+2}\}$;   *(b)* $\{\frac{n+5i-nj}{i+n^3 j-k}\}$;   *(c)* $\{\frac{(1-i+j)^n - k^{n/2}}{n!}\}$;   *(d)* $\{\frac{e^n}{n}\left(1 - \frac{1}{n}i + j\right) + \frac{\cos(n)}{e^n}(j + 7k)^2\}$.

Solution. (a) $0, \infty$;   (b) $0, 0i + 0j + 0k$;   (c) $0, 0i + 0j + 0k$;   (d) $\infty, \infty i + \infty j + \infty k$.

Exercise 3.88. *Let p_0 be a given nonnull quaternion number. Define the sequence $\{p_n\}$ recursively by*

$$p_{n+1} = \frac{1}{2}\left(p_n + \frac{1}{p_n}\right), \quad n \geq 0.$$

If $\{p_n\}$ converges to a quaternion number $p \neq 0_{\mathbb{H}}$, show that: (a) if $\mathrm{Sc}(p_0) > 0$ then $\lim_{n\to\infty} p_n = 1_{\mathbb{H}}$;   (b) if $\mathrm{Sc}(p_0) < 0$ then $\lim_{n\to\infty} p_n = -1_{\mathbb{H}}$;   (c) if $\mathrm{Sc}(p_0) = 0$ and $\mathrm{Vec}(p_0) \neq (0,0,0)$ then $\{p_n\}$ is divergent.

Example. Let $p_n = a_n + b_n i + c_n j + d_n k$, where $a_n, b_n, c_n, d_n \in \mathbb{R}$. Prove that $\lim_{n\to\infty} p_n = p$, $p = a + bi + cj + dk$ $(a, b, c, d \in \mathbb{R})$, if and only if

$$\lim_{n\to\infty} a_n = a, \quad \lim_{n\to\infty} b_n = b, \quad \lim_{n\to\infty} c_n = c, \quad \text{and} \quad \lim_{n\to\infty} d_n = d.$$

Solution. First, we prove the necessity part. Assume that p_n converges to a limit p, i.e. $\lim_{n\to\infty} p_n = p$. Hence, $\lim_{n\to\infty}(p_n - p) = 0_{\mathbb{H}}$. Let $\epsilon > 0$ be arbitrarily chosen and fixed. Then there exists a natural number N such that for every $n > N$ we have

$$\sqrt{(a_n - a)^2 + (b_n - b)^2 + (c_n - c)^2 + (d_n - d)^2} < \epsilon.$$

Since

$$|a_n - a| \leq \sqrt{(a_n - a)^2 + (b_n - b)^2 + (c_n - c)^2 + (d_n - d)^2} < \epsilon,$$

and similarly $|b_n - b|, |c_n - c|, |d_n - d| < \epsilon$, we conclude that

$$\lim_{n\to\infty} a_n = a, \quad \lim_{n\to\infty} b_n = b, \quad \lim_{n\to\infty} c_n = c, \quad \lim_{n\to\infty} d_n = d.$$

For the sufficiency part, assume that the previous relations hold. Then

$$\lim_{n\to\infty} \sqrt{(a_n - a)^2 + (b_n - b)^2 + (c_n - c)^2 + (d_n - d)^2} = 0,$$

i.e. $\lim_{n\to\infty} p_n = p$.

Next let us study the behavior of a quaternion sequence "after a certain point", and give it a precise meaning regardless of what went on before that. This may be done by discarding the first N terms of a given quaternion sequence $\{p_n\}$ to get a shifted sequence $\{q_n\}$ given by $q_n = p_{N+n}$. We shall denote it by $\{p_{N+n}\}$, so that

$$\{p_{N+n}\} = \{p_{N+1}, \; p_{N+2}, \; p_{N+3}, \; p_{N+4}, \; \ldots\}.$$

To complete the reasoning we introduce the definition below.

3.8   Certain 'Property Eventually'

We say that a quaternion sequence $\{p_n\}$ satisfies a 'certain property eventually', if there is a natural number N such that the sequence $\{p_{N+n}\}$ satisfies that property. For instance, a quaternion sequence $\{p_n\}$ is eventually bounded if there exists an N such that the sequence $\{p_{N+n}\}$ is bounded.

Exercise 3.89. *Show that if a quaternion sequence is eventually bounded then it is bounded.*

Exercise 3.90. *Let $\{p_n\} = \{1, i + \frac{1}{2}k, \frac{1}{3}i + \frac{1}{4}k, \frac{1}{5}i + \frac{1}{6}k, \frac{1}{7}i + \frac{1}{8}k, \dots\}$. Prove that the sequence $\{p_n\}$ is eventually bounded.*

We proceed with an example.

Example. Let N be a natural number and $\{p_n\}$ a quaternion sequence. Prove that $\lim_{n\to\infty} p_n = p$ if and only if the shifted sequence $\lim_{n\to\infty} p_{N+n} = p$, $p \in \mathbb{H}$.

Solution. Let $\epsilon > 0$ be arbitrarily chosen and fixed. Then there exists a natural number N_1 such that $|p_n - p| < \epsilon$ whenever $n > N_1$. If $n > N_1$, then $N + n > N_1$, therefore $|p_{N+n} - p| < \epsilon$. Hence, $\lim_{n\to\infty} p_{N+n} = p$. Conversely, suppose that $\lim_{n\to\infty} p_{N+n} = p$. Then there exists a natural number N_2 such that $|p_{N+n} - p| < \epsilon$ whenever $n > N_2$. If $n > N + N_2$, then $n - N > N_2$, so

$$|p_n - p| = |p_{N+(n-N)} - p| < \epsilon.$$

Hence, $\lim_{n\to\infty} p_n = p$.

3.9   Quaternion Subsequence

A quaternion subsequence of $\{p_n\}$ is a sequence of the form $\{p_{n_i}\}$, where $\{n_i\}$ is a strictly increasing sequence of natural numbers.

Exercise 3.91. *Prove that any subsequence of a convergent quaternion sequence is convergent (to the same limit).*

Exercise 3.92. *Prove that every subsequence of a bounded quaternion sequence is bounded.*

Exercise 3.93. *If the shifted subsequence $\{p_{N+n}\}$ is bounded, does it follow that the sequence $\{p_n\}$ is bounded? Justify your answer.*

3.10   Expression 'of the Order of'

If $\{p_n\}$ and $\{q_n\}$ are two quaternion sequences such that a natural number N can be found such that $|(p_n/q_n)| < K$ whenever $n > N$, where K is independent of n, we say that $\{p_n\}$ is 'of the order of' $\{q_n\}$, and write $p_n = O(q_n)$. For example,

$$\left\{ \frac{n + 2j - nk}{n + n^3 i + n^2 k} \right\} = \left\{ \frac{(n + 2j - nk)(n - n^3 i - n^2 k)}{n^2(1 + n^2 + n^4)} \right\} = O\left(\frac{1}{n^2}\right).$$

If $\lim_{n\to\infty}(p_n/q_n) = 0_{\mathbb{H}}$, we write $p_n = o(q_n)$.

Exercise 3.94. *Consider* $\{p_n\} = \left\{\frac{\sqrt{n+1}}{2} + \frac{\sqrt{n+1}}{2}i - \frac{\sqrt{n+1}}{2}j - \frac{\sqrt{n+1}}{2}k\right\}$, *and* $\{q_n\} = \left\{\frac{\sqrt{n}}{2} + \frac{\sqrt{n}}{2}i + \frac{\sqrt{n}}{2}j + \frac{\sqrt{n}}{2}k\right\}$. *Prove that* $p_n = O(q_n)$.

3.11 Equivalent Convergence Criteria

We say that the quaternion sequences $\{p_n\}$ and $\{q_n\}$ are equivalent, and write $p_n \sim q_n$, if $\lim_{n\to\infty} \left|\frac{p_n}{q_n}\right| = 1$.

Exercise 3.95. *Let* $\{p_n\}$ *and* $\{q_n\}$ *be two quaternion sequences such that* $p_n = q_n + o(q_n)$. *Prove that* $p_n \sim q_n$.

Exercise 3.96. *Let* $\{p_n\}$, $\{q_n\}$ *and* $\{r_n\}$ *be quaternion sequences. If* $\{p_n\}$ *and* $\{q_n\}$ *are equivalent, prove that* $\lim_{n\to\infty}(p_n r_n) = \lim_{n\to\infty}(q_n r_n)$.

Despite its many advantages, the definition of convergence of a quaternion sequence has one disadvantage, namely that we need to know the limit in order to prove that the sequence converges. There is a method that permits to prove the existence of a limit even when it cannot be determined explicitly. This test of convergence bears the name of the French mathematician Augustin-Louis Cauchy (1789–1857), and it will be briefly discussed below.

3.12 Cauchy Quaternion Sequence

A quaternion sequence $\{p_n\}$ is said to be a Cauchy quaternion sequence, or fundamental (or to have the Cauchy property) if, for each positive ϵ, no matter how small, an integer $N > 0$ can be found such that $|p_n - p_m| < \epsilon$ for all $n, m > N$. "For all m, n that satisfy $n, m > N$" is essential: for instance, a quaternion sequence $\{p_n\}$ satisfying $|p_{n+1} - p_n| < \epsilon$ for all $n > N$ may not be a Cauchy quaternion sequence.

Example. Consider the quaternion sequence $\{p_n\} = \left\{\frac{1-i+j-k}{2n}\right\}$. Prove that $\{p_n\}$ is a Cauchy quaternion sequence.

Solution. Let $\epsilon > 0$ be arbitrarily chosen and fixed. We have to find a natural number N such that if $n, m > N$, then $|p_n - p_m| < \epsilon$. Note that

$$|p_n - p_m| = \left| \frac{1 - i + j - k}{2n} - \frac{1 - i + j - k}{2m} \right|$$

$$= \left| \frac{1}{n} - \frac{1}{m} \right| \leq \frac{1}{n} + \frac{1}{m}.$$

As in real analysis, if both n and m are greater than $\frac{2}{\epsilon}$, the result will follow.

Exercise 3.97. *Let $\{p_n\}$ and $\{q_n\}$ be Cauchy quaternion sequences. Prove that $\{p_n q_n\}$ is a Cauchy quaternion sequence.*

Exercise 3.98. *Let $p_n = a_n + ib_n + jc_n + kd_n$, where $a_n, b_n, c_n, d_n \in \mathbb{R}$. Prove that $\{p_n\}$ is a Cauchy quaternion sequence if and only if $\{a_n\}, \{b_n\}, \{c_n\}, \{d_n\}$ are Cauchy sequences.*

3.13 Cauchy Convergence Test

A quaternion sequence $\{p_n\}$ is convergent if and only if it is a Cauchy quaternion sequence. That this condition is necessary is trivial: If $\lim_{n \to \infty} p_n = p$, we can find N such that $|p_n - p| < \frac{\epsilon}{2}$. For $n, m > N$, it follows that

$$|p_n - p_m| \leq |p_n - p| + |p_m - p| < \epsilon.$$

There are several ways to prove that the condition is also sufficient. Here we make use of the well-known Bolzano-Weierstrass theorem, attributed to the Czech mathematician and philosopher Bernard Bolzano (1781–1848) and the German mathematician Karl Weierstrass (1815–1897). Suppose that $\{p_n\} = \{a_n\} + \{b_n\}i + \{c_n\}j + \{d_n\}k$ is a Cauchy quaternion sequence. Using the result of the previous exercise, the four real-valued sequences $\{a_n\}, \{b_n\}, \{c_n\}$, and $\{d_n\}$ have the Cauchy property, and moreover they are bounded. In particular, for $\epsilon = 1$ a natural number N can be found such that $n \geq m \geq N$ implies that $|a_n - a_m| < 1$, and so $|a_n| \leq |a_n - a_N| + |a_N| < 1 + |a_N|$. Hence the real sequence $\{a_n\}$ is bounded by $\max\{|a_1|, \ldots, |a_{N-1}|, 1 + |a_N|\}$. By the Bolzano-Weierstrass theorem, $\{a_n\}$ must contain a convergent subsequence, say $\{a_{n_i}\}$. Thus, for some $a \in \mathbb{R}$ we have $\lim_{i \to \infty} a_{n_i} = a$. Let $\epsilon > 0$ be arbitrarily chosen and fixed. Then there exists a natural number I so that $i \geq I$ implies $|a_{n_i} - a| < \frac{\epsilon}{2}$. There is also an integer M such that $n, m \geq M$ we have $|a_n - a_m| < \frac{\epsilon}{2}$. We may also, if necessary, increase I until $n_I \geq M$. Then $n \geq n_I$ implies $|a_n - a| \leq |a_n - a_{n_I}| + |a_{n_I} - a| < \epsilon$. Therefore $\{a_n\}$ converges to a. Similarly we can prove that the sequences $\{b_n\}, \{c_n\}$ and $\{d_n\}$ converge, respectively, to $b, c, d \in \mathbb{R}$. Consequently, $\{p_n\}$ converges to $p = a + bi + cj + dk$.

Exercise 3.99. *Let $\{p_n\}$ be a convergent sequence, and $\{q_n\}$ a Cauchy quaternion sequence. Prove that the quaternion sequence $\{p_n q_n\}$ has the Cauchy property.*

Exercise 3.100. *Let $\{p_n\}$ and $\{q_n\}$ be Cauchy quaternion sequences, and $a, b \in \mathbb{H}$. Prove that: (a) $\{ap_n + bq_n\}$ is a Cauchy sequence;   (b) $\frac{\{p_n\}}{\{q_n\}}$ is a Cauchy sequence whenever $q_n \neq 0_\mathbb{H}$ for all n.*

Let us move on to the discussion of a formula called "summation by parts", which is often used to simplify the computation or (especially) estimation of certain types of sums. The summation by parts formula is often called Abel transformation, named after Niels Henrik Abel (1802–1829).

3.14 Summation by Parts

For any quaternion sequences $\{p_n\}$ and $\{q_n\}$, the following relation holds:

$$\sum_{k=m}^{n}(p_{k+1} - p_k)q_{k+1} + \sum_{k=m}^{n} p_k(q_{k+1} - q_k) = p_{n+1}q_{n+1} - p_m q_m. \tag{3.1}$$

Indeed, combining the two terms on the left-hand side, we formally obtain

$$\sum_{k=m}^{n} (p_{k+1}q_{k+1} - p_k q_{k+1} + p_k q_{k+1} - p_k q_k) = p_{n+1}q_{n+1} - p_m q_m,$$

which is a telescoping sum, leaving us with the term $p_{n+1}q_{n+1} - p_m q_m$.

Exercise 3.101. *Let* $\{p_n\} = \{(n-1)i + (n-1)j\}$ *and* $\{q_n\} = \{(n-1)k\}$. *Find* $\sum_{k=1}^{n}(p_{k+1}q_{k+1} - p_k q_k)$.

Solution. $n^2 i - n^2 j$.

Exercise 3.102. *Let* $\{p_n\} = \{n - i - j\}$ *and* $\{q_n\} = \{n - ni + k\}$. *Find* $\sum_{k=m}^{n}(p_{k+1}q_{k+1} - p_k q_k)$.

Solution. $n^2 + n - m^2 + 2m - (n^2 + 3n + 3 + m^2 + 2m)i - (n+2)j - mk$.

3.15 Abel Transformation

Let $\{p_n\}$ and $\{q_n\}$ be quaternion sequences, and let $\{S_n\} = \{p_1 + p_2 + \cdots + p_n\}$ denote the n-th partial sum of the series corresponding to the sequence $\{p_n\}$. Then for any $m < n$ we have

$$\sum_{k=m+1}^{n} p_k q_k = S_n q_n - S_m q_m - \sum_{k=m}^{n-1} S_k(q_{k+1} - q_k). \tag{3.2}$$

Indeed, applying the summation by parts formula (3.1) to the quaternion sequences $\{S_n\}$ and $\{q_n\}$, we obtain

$$\sum_{k=m}^{n-1}(S_{k+1}-S_k)q_{k+1} + \sum_{k=m}^{n-1}S_k(q_{k+1}-q_k) = S_n q_n - S_m q_m.$$

Since $p_{k+1} = S_{k+1} - S_k$, we conclude that

$$\sum_{k=m}^{n-1}p_{k+1}q_{k+1} + \sum_{k=m}^{n-1}S_k(q_{k+1}-q_k) = S_n q_n - S_m q_m.$$

Replacing k with $k-1$ in the first sum and bringing the second sum to the right-hand side, we obtain (3.2).

Exercise 3.103. *Let $\{p_n\}$ be a convergent quaternion sequence, and $\{q_n\}$ a quaternion sequence such that $\left| \sum_{k=m+1}^{n} p_k q_k \right| < \infty$ for every natural numbers m and n. Prove that $\{q_n\}$ is convergent.*

3.16   Advanced Practical Exercises

1. Let $\{p_n\}$ be a quaternion sequence and $p \in \mathbb{H}$. If $\lim_{n\to\infty}|p_n| = |p|$, does it follow that $\lim_{n\to\infty} p_n = p$? Justify your answer.
2. Let $\{p_n\}$ be a quaternion sequence such that $\lim_{n\to\infty} p_n = p$, $p \in \mathbb{H}$. Prove that $\lim_{n\to\infty} \frac{1}{n}(p_1 + p_2 + \cdots + p_n) = p$.
3. Let $\{p_n\}$ and $\{q_n\}$ be quaternion sequences such that $\lim_{n\to\infty} p_n = p$ and $\lim_{n\to\infty} q_n = q$, for $p, q \in \mathbb{H}$. Let also $a, b \in \mathbb{H}$. Prove that $\lim_{n\to\infty}(p_n a \pm q_n b) = pa \pm qb$.
4. Compute the following limits:

   (a) $\displaystyle \lim_{n\to\infty} \frac{\left(\frac{1}{8} + \frac{1}{2}i + \frac{1}{20}j - \frac{1}{3}k\right)^n}{1 + \left(\frac{1}{8} + \frac{1}{2}i + \frac{1}{20}j - \frac{1}{3}k\right)^{2n}}$;

   (b) $\displaystyle \lim_{n\to\infty} \frac{(1 + i + j + k)^n}{1 + (1 + i + j + k)^{2n}}$;

   (c) $\displaystyle \lim_{n\to\infty}\left(\frac{i - j + k}{1^4} + \frac{(i - j + k)^2}{2^4} + \cdots + \frac{(i - j + k)^n}{n^4}\right)$;

   (d) $\displaystyle \lim_{n\to\infty} \frac{n + i + j + k}{n - i + k}$;

   (e) $\displaystyle \lim_{n\to\infty} \frac{1 + 2n^2 i - j - k}{n^2 + j + k}$;

   (f) $\displaystyle \lim_{n\to\infty} \frac{2 - i + n^2 j - 2k}{1 + n^2 i + j}$;

(g) $\displaystyle\lim_{n\to\infty} \frac{1+i+j+n^2k}{1+n^2k}$;

(h) $\displaystyle\lim_{n\to\infty} \frac{n^3-i+2k}{1-2i+\frac{1}{2}j+k}$;

(i) $\displaystyle\lim_{n\to\infty} \frac{1+ni+nj+nk}{1-i-j-k}$;

(j) $\displaystyle\lim_{n\to\infty} \frac{1+2n^2i-j-k}{n^2+j+k}$;

(k) $\displaystyle\lim_{n\to\infty} \frac{2-i+(n^2+1)j-2k}{1+n^2i+j}$;

(l) $\displaystyle\lim_{n\to\infty} \frac{n+i+j+k}{n-i+2k}$;

(m) $\displaystyle\lim_{n\to\infty} \frac{(n+i+j-k)^2-(n-ni+nj+k)^2}{(n+1)i-j+nk}$.

5. Compute $\displaystyle\lim_{n\to\infty}\left(\sqrt{n+i+j-k}-\sqrt{1-ni+k}\right)$.

6. Compute $\displaystyle\lim_{n\to\infty} \frac{1+i-k+2+2i-2k+\cdots+n+ni-nk}{3n^2-1}$.

7. Compute $\displaystyle\lim_{n\to\infty} \frac{\frac{1}{2}-\frac{1}{4}i-\frac{1}{3}j+k+\cdots+\frac{1}{2}n^2-\frac{1}{4}n^2i-\frac{1}{3}n^2j+n^2k}{n^3}$.

8. Compute the following limits:

(a) $\displaystyle\lim_{n\to\infty}(n+i+j+k)^{\frac{1}{n}}$;

(b) $\displaystyle\lim_{n\to\infty}(1+ni+j+k)^{\frac{1}{n}}$;

(c) $\displaystyle\lim_{n\to\infty}(1+i+nj+k)^{\frac{1}{n}}$;

(d) $\displaystyle\lim_{n\to\infty}(1+i+j+nk)^{\frac{1}{n}}$;

(e) $\displaystyle\lim_{n\to\infty}(n+ni+j+k)^{\frac{1}{n}}$;

(f) $\displaystyle\lim_{n\to\infty}(1-j-k)^{\frac{1}{n}}$;

(g) $\displaystyle\lim_{n\to\infty}(2+k)^{\frac{1}{n}}$.

9. Compute $\displaystyle\lim_{n\to\infty}\left(\frac{(n+2)!+n!}{(n+2)!-n!}+\sum_{l=1}^{n}\frac{1}{\sqrt{n^2+l}}i+\frac{1}{n}\cos\left(\frac{n\pi}{2}\right)j+\left(1+\frac{1}{3n}\right)^n k\right)$.

10. Compute $\displaystyle\lim_{n\to\infty}\left(1+\left(1-\frac{1}{n^2}\right)^n i+\frac{1\cdot3\cdot5\cdots(2n-1)}{2\cdot4\cdot6\cdots2n}j+\frac{1^l+2^l+\cdots+n^l}{n^{l+1}}k\right)$,   for $l\in\mathbb{N}$.

11. Compute

$$\lim_{n\to\infty}\left(n^{\frac{1}{n}}+\frac{1+2+\cdots+n}{n^2}i+\left(\sqrt{2+\sqrt{2+\sqrt{2+\cdots}}}\right)j+\frac{2n-2}{n+1}k\right).$$

12. Compute

(a) $\displaystyle\lim_{n\to\infty}\frac{a_n+2}{a_n+1}$ if $\displaystyle\lim_{n\to\infty}a_n=1-i+j$;

(b) $\lim\limits_{n\to\infty} \dfrac{a_n + 3}{a_n - 4}$ if $\lim\limits_{n\to\infty} a_n = 1 + 2i + k$;

(c) $\lim\limits_{n\to\infty} \dfrac{a_n^2 + i + j}{a_n + k}$ if $\lim\limits_{n\to\infty} a_n = i + j + k$;

(d) $\lim\limits_{n\to\infty} \dfrac{a_n - i - k}{a_n + i + j}$ if $\lim\limits_{n\to\infty} a_n = i - 2j + 3k$;

(e) $\lim\limits_{n\to\infty} \dfrac{a_n + i + j + k}{a_n}$ if $\lim\limits_{n\to\infty} a_n = 3 - i + k$;

(f) $\lim\limits_{n\to\infty} \dfrac{a_n - 2i + k}{a_n + 1}$ if $\lim\limits_{n\to\infty} a_n = 2$;

(g) $\lim\limits_{n\to\infty} \dfrac{a_n^2 + 4i + 5k}{a_n^3 + 3k}$ if $\lim\limits_{n\to\infty} a_n = 5$;

(h) $\lim\limits_{n\to\infty} \dfrac{a_n - i - j}{a_n + 2k}$ if $\lim\limits_{n\to\infty} a_n = 3$;

(i) $\lim\limits_{n\to\infty} \dfrac{a_n^2 - i - j + 3k}{a_n + i + j}$ if $\lim\limits_{n\to\infty} a_n = i + k$;

(j) $\lim\limits_{n\to\infty} \dfrac{a_n - i - k}{a_n + j + k}$ if $\lim\limits_{n\to\infty} a_n = i + j + k$.

13. Let $\{a_n\}$ be a quaternion sequence such that $\lim_{n\to\infty}\left|\dfrac{a_{n+1}}{a_n}\right| > 1$. Prove that $\lim_{n\to\infty} \dfrac{a_n}{a_1} = \pm\infty$.

14. Let $\{a_n\}$ be a quaternion sequence such that $\left|\lim_{n\to\infty} \dfrac{a_{n+1}}{a_n}\right| < 1$. Prove that $\lim_{n\to\infty} a_n = 0$.

15. Compute

   (a) $\lim_{n\to\infty}(a_n + b_n i + c_n j)$, where $a_n = \sum_{l=1}^{n} \dfrac{n}{n^2+l}$, $b_n = \sum_{l=1}^{n} \dfrac{1}{\sqrt{n^2+l}}$, and
   $$c_n = \sum_{l=1}^{n} \left(\dfrac{n}{n^4+l}\right)^{\frac{1}{3}};$$

   (b) $\lim_{n\to\infty}(a_n j + b_n k)$, where $a_n = \sum_{l=1}^{n} \left(\sqrt{1 + \dfrac{l}{n^2}} - 1\right)$, and
   $$b_n = \sum_{l=1}^{n} \left(\left(1 + \dfrac{l^2}{n^3}\right)^{\frac{1}{3}} - 1\right).$$

16. Find the limit of the quaternion sequence $\{p_n\}$ if

   (a) $p_{n+1} = i p_n + j$;

   (b) $j p_{n+1} = k p_n - i$;

   (c) $(1 + i) p_{n+1} = k p_n$;

   (d) $p_{n+1} k = (1 - i - j) p_n + k$;

   (e) $(1 - i - j) p_{n+1} = k p_n - 1 + i$;

   (f) $p_{n+1}^2 = i p_n + p_{n-1}$;

   (g) $i p_{n-1} = (i + j) p_n^2$.

17. Prove that $p_n = O(q_n)$, where

   (a) $p_n = \dfrac{\sqrt{2n^2+n-2}}{2} + \dfrac{1}{2}i + \dfrac{1}{2}j + \dfrac{\sqrt{2n^2+n+4}}{2}k$ and $q_n = \dfrac{\sqrt{n^2-3}}{2} + \dfrac{\sqrt{n^2+3}}{2}i + \dfrac{\sqrt{n^2-1}}{2}j + \dfrac{\sqrt{n^2+2}}{2}k$;

$\quad$ (b) $p_n = \dfrac{\sqrt{2^{n+1}+3^{n+1}-3}}{2} + \dfrac{\sqrt{2^{n+1}+3}}{2}i + \dfrac{\sqrt{3^{n+1}-3}}{2}j - \dfrac{\sqrt{2^{n+1}+3}}{2}k$ and $q_n = \dfrac{\sqrt{2^n+3}}{2} - \dfrac{\sqrt{3^n-3}}{2}i + \dfrac{\sqrt{2^n+3^n-3}}{2}j + \dfrac{\sqrt{2^n+3}}{2}k.$

18. Prove that $p_n = o(q_n)$, where

$\quad$ (a) $p_n = \sqrt{2^n + 3^n - 2} + \sqrt{2^n - 2}\,i + \sqrt{3^n + 2}\,j + \sqrt{2^n + 3^n + 2}\,k$, and $q_n = \sqrt{4^n - 1} - \sqrt{4^n + 1}\,i + \sqrt{4^n - 2}\,j - \sqrt{4^n + 2}\,k$;

$\quad$ (b) $p_n = \sqrt{n\cos n! + 2} - \sqrt{n\cos n! - 3}\,i - \sqrt{n\cos n! - 2}\,j - \sqrt{n\cos n! + 2}\,k$, and $q_n = \sqrt{n^2 + 2} + \sqrt{n^2 - 2}\,i + j + k$.

19. Let $\{p_n\}$ be a quaternion sequence. Prove that $p_n \sim p_n$.

20. Let $\{p_n\}$ and $\{q_n\}$ be two quaternion sequences such that $p_n \sim q_n$. Prove that $q_n \sim p_n$.

21. Let $\{p_n\}$, $\{q_n\}$ and $\{r_n\}$ be quaternion sequences such that $p_n \sim q_n$ and $q_n \sim r_n$. Prove that $p_n \sim r_n$.

22. Let $a \in \mathbb{R}^+$, $k \in \mathbb{N}$ and $s_n = \min\{m \in \mathbb{N} : a \leq \left(\frac{m}{n}\right)^k\}$. Prove that $p_n = 1 + i + \frac{s_n}{n}j + \sqrt{\frac{s_n}{n}}k$ is a Cauchy quaternion sequence.

23. Let s_n be as in the previous exercise, and $b > 1$, $c > 0$, $n \in \mathbb{N}$, $t_n = \min\{g \in \mathbb{Z} : c^n \leq b^g\}$. Prove that $p_n = 1 + \frac{s_n}{n}i + \frac{t_n}{n}j + \frac{t_n^2}{n^2}k$ is a Cauchy quaternion sequence.

24. Let $f_n \in \mathbb{R}$ and $f_{n+1} = \frac{r}{1+f_n^2} + d$, $|r| < 1$, $d \in \mathbb{R}$, and let s_n and t_n be as in the previous exercises. Prove that $p_n = 1 + \frac{s_n}{n}i + \frac{t_n}{n}j + f_n k$ is a Cauchy quaternion sequence.

25. Let $p_n, w_n \in \mathbb{H}$, $\{p_n\}$ a convergent quaternion sequence, and $\{w_n\}$ a Cauchy quaternion sequence. Prove that $\{p_n w_n\}$ has the Cauchy property.

Quaternion Series and Infinite Products

4

An essential feature of the classical theory of power series is that we can manipulate recurrence relations for power series without necessarily worrying about whether the underlying series converge. In case they do converge, we can extract important information about the recurrence relation that may not otherwise be easily obtainable.

In this chapter we use the properties of quaternion sequences described in the previous chapter to explore the key notions of convergence and divergence for quaternion infinite series and products. In practice it is truly laborious and frequently difficult to prove that a particular quaternion series converges uniformly in a given region. Here we preferably end up dealing (implicitly or explicitly) with general results that provide some tests under which quaternion series will converge uniformly. As we will see, many interesting and useful quaternion series and products are simply generalizations of their counterparts from real and complex calculus. The reader is urged to pay particular attention to what is said about geometric series, since this type of series will be important in the later chapters.

We now introduce the notion of a quaternion series in a manner similar to that encountered in real and complex calculus.

4.1 Quaternion Series

Let $\{p_n\}$ be a sequence of real quaternions, and let the sum $p_1 + p_2 + \cdots + p_n$ be denoted by S_n. The sequence $\{S_n\}$ is also called the series associated to the sequence $\{p_n\}$.

We introduce the following definitions:

(i) If S_n tends to a limit S as $n \to \infty$, the quaternion infinite series $\sum_{n=1}^{\infty} p_n$ is said to be convergent, or to converge to the sum S, and we write $\sum_{n=1}^{\infty} p_n = S$;

(ii) $\sum_{n=1}^{\infty} p_n$ is said to diverge if $\{S_n\}$ does not converge as $n \to \infty$;

(iii) $\sum_{n=1}^{\infty} p_n$ is said to diverge to infinity if $\{S_n\}$ tends to infinity as $n \to \infty$;

(iv) $\sum_{n=1}^{\infty} p_n$ is said to diverges to negative infinity if $\{S_n\}$ tends to negative infinity as $n \to \infty$.

J.P. Morais et al., *Real Quaternionic Calculus Handbook*,
DOI 10.1007/978-3-0348-0622-0_4, © Springer Basel 2014

When the underlying quaternion series converges, the expression $S - S_n$, which is the sum of the series

$$p_{n+1} + p_{n+2} + p_{n+3} + \cdots,$$

is called the remainder after n terms, and is frequently denoted by the symbol R_n. In addition, the sum $p_{n+1} + p_{n+2} + p_{n+3} + \cdots + p_{n+m}$ will be denoted by $S_{n,m}$.

We can easily derive some properties for quaternion series:

4.2   Arithmetic Properties of Quaternion Series

Let $\sum_{n=1}^{\infty} p_n$ and $\sum_{n=1}^{\infty} q_n$ be two quaternion series, and λ, μ quaternion numbers. The following statements are true:

(i) If $\sum_{n=1}^{\infty} p_n$ and $\sum_{n=1}^{\infty} q_n$ are both convergent, then the series $\sum_{n=1}^{\infty} (\lambda p_n \pm \mu q_n)$ are convergent and their sum (difference) is $\sum_{n=1}^{\infty} (\lambda p_n \pm \mu q_n) = \lambda \sum_{n=1}^{\infty} p_n \pm \mu \sum_{n=1}^{\infty} q_n$;

(ii) If $\sum_{n=1}^{\infty} p_n$ is convergent and $\sum_{n=1}^{\infty} q_n$ is divergent, then the series $\sum_{n=1}^{\infty} (p_n \pm q_n)$ are divergent;

(iii) If N is a natural number, then the quaternion series $\sum_{n=1}^{\infty} p_n$ converges if and only if the shifted series $\sum_{n=1}^{\infty} p_{N+n}$ converges.

Exercise 4.104. *Let $\{p_n\} = \{1 + ni + j + k\}$ and $\{q_n\} = \{\frac{1}{n^2} i + j\}$. Investigate the convergence or divergence of the series $\sum_{n=1}^{\infty} (p_n \pm q_n)$.*

Solution. Both divergent.

4.3   Geometric Quaternion Series

Due to the noncommutativity of the multiplication we shall distinguish between a left and right geometric quaternion series. A left geometric quaternion series is any series of the form

$$\sum_{n=1}^{\infty} ap^{n-1} = a + ap + ap^2 + ap^3 + \cdots, \tag{4.1}$$

where $a \in \mathbb{H}$. Similarly, we define a right geometric quaternion series as

$$\sum_{n=1}^{\infty} p^{n-1}a = a + pa + p^2a + p^3a + \cdots.$$

Throughout the text we only use left geometric quaternion series that, for simplicity, we call geometric quaternion series. Nevertheless, all results obtained for left geometric quaternion series can easily be adapted to right geometric quaternion series. For (4.1), the nth term of the sequence of partial sums is

$$S_n = a + ap + ap^2 + \cdots + ap^{n-1}.$$

In order to find a formula for S_n, we multiply the right-hand side of S_n by p,

$$S_n p = ap + ap^2 + ap^3 + \cdots + ap^n,$$

and subtract this result from S_n. Then all terms cancel, except for the first term in S_n and the last term in $S_n p$. So, more concretely,

$$S_n - S_n p = a + ap + ap^2 + \cdots + ap^{n-1} - \left(ap + ap^2 + \cdots + ap^{n-1} + ap^n\right)$$

$$= a - ap^n,$$

or $S_n(1_{\mathbb{H}} - p) = a(1_{\mathbb{H}} - p^n)$. Solving the last equation for S_n one obtains

$$S_n = a(1_{\mathbb{H}} - p^n)(1_{\mathbb{H}} - p)^{-1}. \tag{4.2}$$

Since $p^n \to 0_{\mathbb{H}}$ as $n \to \infty$ whenever $|p| < 1$, we have that $S_n \to a(1_{\mathbb{H}} - p)^{-1}$. In other words, for $|p| < 1$ the sum of a geometric series (4.1) is

$$a(1_{\mathbb{H}} - p)^{-1} = a + ap + ap^2 + ap^3 + \cdots . \tag{4.3}$$

The geometric series (4.1) diverges when $|p| \geq 1$. Let us give an example. Consider the quaternion geometric series

$$\sum_{n=1}^{\infty} \frac{(1 - i + 2j + k)^n}{3^n}.$$

We see that it has the form given in (4.1), with $a = \frac{1}{3}(1 - i + 2j + k)$ and $p = \frac{1}{3}(1 - i + 2j + k)$. Since $|p| = \frac{\sqrt{7}}{3} < 1$, the series is convergent and its sum is given by (4.3):

$$\sum_{n=1}^{\infty} \frac{(1 - i + 2j + k)^n}{3^n} = \frac{(1 - i + 2j + k)}{3} \left(1_{\mathbb{H}} - \frac{(1 - i + 2j + k)}{3}\right)^{-1}$$

$$= -\frac{2}{5} - \frac{3}{10}i + \frac{3}{5}j + \frac{3}{10}k.$$

Exercise 4.105. *Use the sequence of partial sums to show that the following series is convergent:* (a) $\sum_{m=1}^{\infty} \frac{1+i-k}{3^{m-1}}$;   (b) $\sum_{m=1}^{\infty} \frac{1}{(i+2-3k)^m}$.

Setting $a = 1_{\mathbb{H}}$, the equality in (4.3) becomes

$$(1_{\mathbb{H}} - p)^{-1} = 1_{\mathbb{H}} + p + p^2 + p^3 + \cdots . \qquad (4.4)$$

Replacing the symbol p by $-p$ in the previous equality, we get a similar result:

$$(1_{\mathbb{H}} + p)^{-1} = 1_{\mathbb{H}} - p + p^2 - p^3 + \cdots .$$

Analogous to (4.3), the last equality is also valid for $|p| < 1$ since $|-p| = |p|$. Now let $a = 1_{\mathbb{H}}$. That is to say, (4.2) gives us the sum of the first n terms of the quaternion series in (4.4):

$$(1_{\mathbb{H}} - p^n)(1_{\mathbb{H}} - p)^{-1} = 1_{\mathbb{H}} + p + p^2 + p^3 + \cdots + p^{n-1}.$$

Ultimately, we obtain the alternative form

$$(1_{\mathbb{H}} - p)^{-1} = 1_{\mathbb{H}} + p + p^2 + p^3 + \cdots + p^{n-1} + p^n(1_{\mathbb{H}} - p)^{-1}.$$

Exercise 4.106. *Let $p = \frac{1}{3}i + \frac{1}{3}j$. Compute $(1_{\mathbb{H}} - p)^{-1}$.*

Solution. Since $|p| = \frac{\sqrt{2}}{3} < 1$, we have $(1_{\mathbb{H}} - p)^{-1} = 1_{\mathbb{H}} + \frac{1}{3}i + \frac{1}{3}j + (\frac{1}{3}i + \frac{1}{3}j)^2 + \cdots .$

We proceed by discussing important results about convergence and divergence of infinite quaternion series. We advice the reader that, while studying the convergence or divergence of a given quaternion series, he should also state by which test you are drawing your conclusions.

4.4   A Necessary Condition for Convergence

If the series $\sum_{n=1}^{\infty} p_n$ converges, then $\lim_{n \to \infty} p_n = 0_{\mathbb{H}}$. Indeed, let S denote the sum of the series. Then $S_n \to S$ and $S_{n-1} \to S$ as n tends to infinity. By taking the limit of both sides of $S_n - S_{n-1} = p_n$ as $n \to \infty$ we obtain the desired conclusion.

Example. The condition above is necessary, but not sufficient: it is possible that $\lim_{n \to \infty} p_n = 0_{\mathbb{H}}$, but $\sum_{n=1}^{\infty} p_n$ diverges. Take for example

$$\{p_n\} = \left\{ \frac{1}{n} + \frac{1}{n}i + \frac{1}{n}j + \frac{1}{n}k \right\} .$$

It is easily seen that $\lim_{n \to \infty} p_n = 0_{\mathbb{H}}$, but $\sum_{n=1}^{\infty} p_n$ is divergent.

4.5   The nth Term Test for Divergence

If $\lim_{n\to\infty} p_n \neq 0_{\mathbb{H}}$ or if the limit does not exist, then $\sum_{n=1}^{\infty} p_n$ diverges. For example, the quaternion series

$$\sum_{n=1}^{\infty} \frac{n^2 + i + n^2 j - nk}{n^2}$$

diverges since, obviously,

$$\lim_{n\to\infty} \frac{n^2 + i + n^2 j - nk}{n^2} = 1 + j \neq 0_{\mathbb{H}}.$$

Nevertheless, the term test cannot prove by itself that a quaternion series converges. We saw above that even if $\lim_{n\to\infty} p_n = 0_{\mathbb{H}}$ the series $\sum_{n=1}^{\infty} p_n$ may or may not converge. In other words, if $\lim_{n\to\infty} p_n = 0_{\mathbb{H}}$, the test is inconclusive; the reader must use another test to determine the convergence or divergence of the series.

Exercise 4.107. *Let $\{p_n\} = \{\frac{1}{n^l} + \frac{1}{n^l}i + \frac{1}{n^l}j + \frac{1}{n^l}k\}$. Prove that the series $\sum_{n=1}^{\infty} p_n$ is divergent whenever $l \in (0, 1)$.*

4.6   Absolute and Conditional Convergence of a Quaternion Series

An infinite quaternion series $\sum_{n=1}^{\infty} p_n$ is said to be absolutely convergent if the series $\sum_{n=1}^{\infty} |p_n|$ converges. An infinite series $\sum_{n=1}^{\infty} p_n$ is said to be conditionally convergent if it converges, but $\sum_{n=1}^{\infty} |p_n|$ diverges. We further assume the reader to be familiar with the fact if $|p_n| \leq C|q_n|$ for all n (C is some number independent of n), then
 (i) If $\sum_{n=1}^{\infty} |q_n|$ converges, then $\sum_{n=1}^{\infty} |p_n|$ converges;
(ii) If $\sum_{n=1}^{\infty} |p_n|$ diverges, then $\sum_{n=1}^{\infty} |q_n|$ diverges.

Example. Consider the quaternion series $\sum_{n=1}^{\infty} \frac{(-1)^n}{2^n} \frac{(1-i-j+k)^n}{n^2+3}$. A direct computation shows that

$$\sum_{n=1}^{\infty} \left| \frac{(-1)^n}{2^n} \frac{(1 - i - j + k)^n}{n^2 + 3} \right| = \sum_{n=1}^{\infty} \frac{1}{n^2 + 3} \leq \sum_{n=1}^{\infty} \frac{1}{n^2}.$$

Since $\sum_{n=1}^{\infty} \frac{1}{n^2}$ is a convergent series, according to the above discussion we conclude that the given series is absolutely convergent.

Exercise 4.108. *Let $\{p_n\} = \{\frac{1}{\sqrt{10}^n} \frac{(1+2i-2j+k)^n}{\sqrt{n^7+1}}\}$. Prove that $\sum_{n=1}^{\infty} p_n$ is absolutely convergent.*

Exercise 4.109. *Let* $\{p_n\} = \{\frac{(-1)^n}{\sqrt{n}}(i - j) - \frac{1}{3^n}k\}$. *Does the series* $\sum_{n=1}^{\infty} p_n$ *converge? Does it converge absolutely? Justify your answer.*

Solution. It converges, but does not converge absolutely.

Exercise 4.110. *Let* $\{p_n\} = \{(-1)^n \frac{1}{n}\}$. *Does* $\sum_{n=1}^{\infty} p_n$ *converge absolutely, converge conditionally, or diverge? Justify your answer.*

Solution. It converges conditionally.

Next we introduce two of the most frequently used tests for convergence of infinite quaternion series.

4.7   Root Test

Suppose $\sum_{n=1}^{\infty} p_n$ is a quaternion series such that

$$\limsup_{n \to \infty} \sqrt[n]{|p_n|} = L_1. \tag{4.5}$$

  (i) If $L_1 < 1$, the series converges absolutely;
 (ii) If $L_1 > 1$ or $L_1 = \infty$, the series diverges;
(iii) If $L_1 = 1$, the test is inconclusive.

For simplicity we shall prove Statements (i) and (ii) only. The verification of Statement (iii) is left to the reader. To prove (i) we take a real number l_1 such that $L_1 < l_1 < 1$. Then a natural number N can be found such that $\sqrt[n]{|p_n|} < l_1$ for $n \geq N$. Having chosen such a l_1, the terms of the series $\sum_{n=1}^{\infty} |p_n|$ are bounded from above by the terms of the real geometric series $\sum_{n=1}^{\infty} l_1^n$, and this means that our series converges absolutely. If we have in (ii) $L_1 > 1$, then we have infinitely many indices n such that $\sqrt[n]{|p_n|} > 1$ (resp. $|p_n| > 1$), implying the divergence of the series.

Example. Let $\{p_n\} = \{e^{-n} + e^{-2n}i + e^{-3n}j + e^{-4n}k\}$. Study the convergence of the series $\sum_{n=1}^{\infty} p_n$ using the root test.

Solution. A direct computation shows that

$$\limsup_{n \to \infty} \sqrt[n]{|p_n|} < 2e^{-2} < 1.$$

Hence, $\sum_{n=1}^{\infty} p_n$ converges absolutely.

Exercise 4.111. *Using the root test, determine whether the quaternion series* $\sum_{n=1}^{\infty} \frac{(3+11i-j+k)^n}{n!}$ *is convergent or divergent.*

Solution. Divergent.

4.8 Ratio Test

Suppose $\sum_{n=1}^{\infty} p_n$ is a series of nonzero quaternion terms such that

$$\limsup_{n\to\infty} \frac{|p_{n+1}|}{|p_n|} = L_2. \tag{4.6}$$

(i) For the L_1 from the root test we have $L_1 \le L_2$, and if $L_2 < 1$ the series
   converges absolutely;
(ii) If $L_2 > 1$ or $L_2 = \infty$, the series diverges;
(iii) If $L_2 = 1$, the test is inconclusive.

We notice that instead of writing $\frac{|p_{n+1}|}{|p_n|}$ we may also take the quotient $\left|\frac{p_{n+1}}{p_n}\right|$. It is
assumed above that the terms are different from zero. Take $L_2 < 1$, and then choose
$L_2 < l_2 < 1$. For $n \ge N$ with a suitable N we have that

$$|p_n| \le l_2|p_{n-1}| \le l_2^2|p_{n-2}| \le \cdots \le l_2^{n-N}|p_N|$$

and, moreover, $|p_n| \le \frac{l_2^n|p_N|}{l_2^N}$, whence $\sqrt[n]{|p_n|} \le l_2 \sqrt[n]{\frac{|p_N|}{l_2^N}}$.

From the last inequality it follows that $L_1 \le L_2$, and from the first we see that,
up to a factor, our series is bounded above by $\sum_{n=1}^{\infty} l_2^n$. And, by an argument such
as was just used above, we have convergence of the series. For the second part of
the proof take $L_2 > l_2 > 1$. Then $|p_{n+1}| \ge l_2|p_n| \ge |p_n|$ for all sufficiently large n.
Hence the absolute values of the generic terms of our series do not converge to zero
and the series diverges.

Example. Let $\{p_n\} = \{(1 + i + 3j + k)^n\}$. Study the convergence of the series
$\sum_{n=1}^{\infty} p_n$ using the ratio test.

Solution. A direct computation shows that

$$\limsup_{n\to\infty} \frac{|p_{n+1}|}{|p_n|} = \sup|1 + i + 3j + k| = 2\sqrt{3} > 1.$$

Therefore the series $\sum_{n=1}^{\infty} p_n$ is divergent.

Exercise 4.112. *Using the ratio test, determine whether the quaternion series*
$\sum_{n=1}^{\infty} \frac{(3-2i+k)^n}{n3^n}$ *is convergent or divergent.*

Solution. Divergent.

4.9   Quaternion Sequences of Bounded Variation

A quaternion sequence $\{p_n\}$ is said to be of bounded variation if $\sum_{n=1}^{\infty} |p_{n+1} - p_n| < \infty$. An important property is that any quaternion sequence of bounded variation converges. Indeed, assume that $\{p_n\}$ is of bounded variation. Given $m < n$, it suffices to observe that $p_n - p_m$ can be written as $\sum_{k=m}^{n}(p_{k+1} - p_k)$. It follows that $|p_n - p_m| \leq \sum_{k=m}^{n} |p_{k+1} - p_k|$. By assumption, the sum $\sum_{k=1}^{\infty} |p_{k+1} - p_k|$ converges, so the sum on the right-hand side of this inequality can be made arbitrarily small as $m, n \to \infty$. In conclusion, $\{p_n\}$ has the Cauchy property and hence converges.

Exercise 4.113. *Let* $\{p_n\} = \left\{ \frac{1}{n^2} + \frac{1}{(n-1)^2}i + \frac{1}{(n-1)^3}j \right\}$. *Prove that* $\{p_n\}$ *is a sequence of bounded variation.*

Exercise 4.114. *Show that if a quaternion sequence* $\{p_n\}$ *is of bounded variation, then* $\lim_{n \to \infty} p_n$ *exists.*

4.10   Dirichlet's Test

Let $\{p_n\}$ be a sequence of quaternion numbers. Suppose that the partial sums of the series $\sum_{n=1}^{\infty} p_n$ are uniformly bounded (although the series $\sum_{n=1}^{\infty} p_n$ may not converge). Then for any quaternion sequence $\{q_n\}$ of bounded variation that converges to zero, the series $\sum_{n=1}^{\infty} p_n q_n$ converges. To show this, define $p_0 = 0_{\mathbb{H}}$ (so that $S_0 = 0_{\mathbb{H}}$) and $q_0 = 0_{\mathbb{H}}$. Setting $m = 0$ in the Abel transformation (3.2), we can write

$$\sum_{k=1}^{n} p_k q_k = S_n q_n - \sum_{k=1}^{n-1} S_k(q_{k+1} - q_k). \tag{4.7}$$

We have two assumptions: the first is that the partial sums $\{S_n\}$ are bounded, say by a constant C, and the second is that the sequence $\{q_n\}$ is of bounded variation and converges to zero. Since $\{S_n\}$ is bounded and $q_n \to 0_{\mathbb{H}}$ as $n \to \infty$, it follows that $S_n q_n \to 0_{\mathbb{H}}$ as $n \to \infty$. Since $|S_n| \leq C$ for all n and $\{q_n\}$ is of bounded variation, the series $\sum_{k=1}^{\infty} S_k(q_{k+1} - q_k)$ is absolutely convergent. Indeed,

$$\sum_{k=1}^{\infty} |S_k(q_{k+1} - q_k)| \leq C \sum_{k=1}^{\infty} |q_{k+1} - q_k| < \infty.$$

Finally, taking $n \to \infty$ in (4.7) it follows that the series $\sum_{k=1}^{\infty} p_k q_k$ converges and its sum equals $\sum_{k=1}^{\infty} S_k(q_{k+1} - q_k)$.

Example. Let $p_n = \frac{(-1)^n}{n^2}$ and $q_n = \frac{1}{n^2-1}i - \sin(\frac{1}{n})j$. Using Dirichlet's test, prove that $\sum_{n=2}^{\infty} p_n q_n$ is convergent.

Solution. It is easy to see that the partial sums of the series $\sum_{n=1}^{\infty} p_n$ are uniformly bounded, and $\{q_n\}$ is of bounded variation and converges to zero. By Dirichlet's test, $\sum_{n=1}^{\infty} p_n q_n$ converges.

Exercise 4.115. *Let $p_n = \{\frac{1}{n^2} + (-1)^n \frac{1}{n}i\}$ and $q_n = \{\frac{1}{n^2+1}i + \frac{1}{n^2+2}k\}$. Show that $\sum_{n=1}^{\infty} p_n q_n$ is convergent.*

4.11 Alternating Series Test

If $\{p_n\}$ is a quaternion sequence of bounded variation that converges to zero, then the series $\sum_{n=1}^{\infty}(-1)^n p_n$ converges. Indeed, since the partial sums of $\sum_{n=1}^{\infty}(-1)^n$ are bounded and $\{p_n\}$ is of bounded variation and converges to zero, the series $\sum_{n=1}^{\infty}(-1)^n p_n$ converges by the Dirichlet test.

Exercise 4.116. *Prove that the series $\sum_{n=1}^{\infty}(-1)^n \left(\frac{1}{n^2}i - \frac{1}{n^2+1}j + \frac{1}{n^2+2}k \right)$ and $\sum_{n=1}^{\infty}(-1)^n \left(\frac{1}{n}i - \frac{1}{\sqrt{n^7+1}}j + \frac{1}{n^4+1}k \right)$ are convergent.*

Exercise 4.117. *Prove that the series $\sum_{n=1}^{\infty} \left(\frac{(-1)^n}{n} + \frac{1}{n}i + \frac{(-1)^n}{n^2+1}j \right)$ is divergent.*

4.12 Abel's Test

Let $\{p_n\}$ be a sequence of quaternion numbers. Suppose that $\sum_{n=1}^{\infty} p_n$ converges. Then for any quaternion sequence $\{q_n\}$ of bounded variation, the series $\sum_{n=1}^{\infty} p_n q_n$ converges.

By Abel's transformation (3.2), for $m < n$ one has that

$$\sum_{k=m+1}^{n} p_k q_k = S_n q_n - S_m q_m - \sum_{k=m}^{n-1} S_k(q_{k+1} - q_k), \qquad (4.8)$$

where S_n denotes the n-th partial sum of the series $\sum_{n=1}^{\infty} p_n$. Let $\sum_{n=1}^{\infty} p_n = S$.
Note that

$$\sum_{k=m}^{n-1} S_k(q_{k+1} - q_k) = \sum_{k=m}^{n-1}(S_k - S)(q_{k+1} - q_k) + S\sum_{k=m}^{n-1}(q_{k+1} - q_k)$$

$$= \sum_{k=m}^{n-1}(S_k - S)(q_{k+1} - q_k) + S q_n - S q_m.$$

Substituting this into (4.8) we conclude that

$$\sum_{k=m+1}^{n} p_k q_k = (S_n - S)q_n - (S_m - S)q_m - \sum_{k=m}^{n-1}(S_k - S)(q_{k+1} - q_k).$$

Let $\epsilon > 0$. Since the sequence $\{q_n\}$ is of bounded variation, it converges. Since $S_n \to S$ as $n \to \infty$, it follows in particular that $(S_n - S)q_n \to 0_{\mathbb{H}}$ and $(S_m - S)q_m \to 0_{\mathbb{H}}$. Thus, we can choose N such that for $n, m > N$, we have $|(S_n - S)q_n| < \epsilon/3$, $|(S_m - S)q_m| < \epsilon/3$, and $|S_n - S| < \epsilon/3$. Thus, for $N < m < n$, we get

$$\left| \sum_{k=m+1}^{n} p_k q_k \right| \leq |(S_n - S)q_n| + |(S_m - S)q_m|$$

$$+ \sum_{k=m}^{n-1} |(S_k - S)(q_{k+1} - q_k)|$$

$$= \frac{\epsilon}{3} + \frac{\epsilon}{3} + \frac{\epsilon}{3} \sum_{k=m}^{n-1} |q_{k+1} - q_k|.$$

Finally, since $\sum_{n=1}^{\infty} |q_{n+1} - q_n|$ converges, the sum $\sum_{k=m}^{n-1} |q_{k+1} - q_k|$ can be made less than 1 for N chosen larger if necessary. In conclusion, for $N < m < n$ we have $\left| \sum_{k=m+1}^{n} p_k q_k \right| < \epsilon$.

Exercise 4.118. *Let $\{p_n\}$, $\{q_n\}$ be sequences of quaternion numbers and assume that $p_n \to 0_{\mathbb{H}}$ and $\frac{1}{n} \sum_{k=1}^{n} k|q_{k+1} - q_k| \to 0_{\mathbb{H}}$ as $n \to \infty$, and that for some real constant C, we have $\left| \frac{1}{n} \sum_{k=1}^{n} p_k \right| \leq C$ for all n. Prove that $\frac{1}{n} \sum_{k=1}^{n} p_k q_k \to 0_{\mathbb{H}}$ as $n \to \infty$.*

We continue our investigation of quaternion power series, and provide some tests that determine whether a series is convergent or divergent.

4.13 Quaternion Power Series

An infinite series of the form

$$\sum_{n=0}^{\infty} a_n(p - q)^n = a_0 + a_1(p - q) + a_2(p - q)^2 + \cdots \tag{4.9}$$

where the coefficients a_n are quaternion constants, is called a left quaternion power series in $p - q$. Analogously, we define a (right) quaternion power series by

$$\sum_{n=0}^{\infty}(p-q)^n a_n = a_0 + (p-q)a_1 + (p-q)^2 a_2 + \cdots. \qquad (4.10)$$

The power series (4.9) and (4.10) are said to be centered at q; the quaternion point q is referred to as the center of the series. It is also convenient to define $(p-q)^0 = 1_{\mathbb{H}}$ even when $p = q$. From now on we only deal with the series (4.9), which for simplicity we call quaternion power series.

Exercise 4.119. *Suppose that $p^N - q^N = \sum_{n=1}^{N} a_n (p-q)^n$ holds for all $p \in \mathbb{H}$ and N a positive integer. (a) Show that $a_n = \binom{N}{n}q^{N-n}$; (b) Replace in the above p by the symbol $p + q$ and show that $(p+q)^N = \sum_{n=0}^{N} \binom{N}{n} p^n q^{N-n}$.*

4.14   Radius of Convergence

There exists a number R, $0 \le R \le +\infty$, called the radius of convergence of the series (4.9), such that the series converges for all values of p such that $|p - q| < R$ and diverges for all p such that $|p - q| > R$. On the boundary, that is, where $|p - q| = R$, the situation is more delicate, as one may have either convergence or divergence (see the example below). In particular, the power series converges uniformly in every smaller ball $\{|p - q| \le R - \epsilon : \epsilon > 0\}$. We notice, for future reference, that the number R may be expressed in terms of the quaternion sequence $\{a_n\}$ of coefficients of the series as

$$\frac{1}{R} := \lim_{n \to \infty} \sup |a_n|^{1/n},$$

with the identifications $\frac{1}{0} := \infty$ and $\frac{1}{\infty} := 0$. This is known as the Cauchy-Hadamard formula for the radius of convergence. The radius of convergence can be:
  (i) $R = 0$ (in which case (4.9) converges only at its center $p = q$), or
  (ii) R is a finite positive number (in which case (4.9) converges at all interior points of the ball $|p - q| < R$), or
(iii) $R = \infty$ (in which case (4.9) converges for all p).

   For simplicity, we assume that the series is centered at $q = 0_{\mathbb{H}}$. Let R be given as above, and assume firstly $0 < R < \infty$. From the definition of R it follows that for all $\epsilon > 0$ there exists an integer $N(\epsilon) > 0$, such that $|a_n|^{1/n}(R - \epsilon) < 1$ for all $n > N(\epsilon)$. Hence, for these n's one has that $|a_n p^n| \le \left(\frac{|p|}{R-\epsilon}\right)^n$. The right-hand side is the form of a convergent series for $|p| \le R - 2\epsilon$, and by comparison we conclude the uniform convergence of the power series for $|p| \le R - 2\epsilon$. Since $\epsilon > 0$ was chosen arbitrarily, we have convergence for $|p| < R$. To prove the divergence for

$|p| > R$, we note that we have $|a_n|^{1/n}(R + \epsilon) \geq 1$ for infinitely many $n \in \mathbb{N}$. Therefore, we find $|a_n||p|^n \geq \left(\frac{|p|}{R+\epsilon}\right)^n$. As we can find for every $|p| > R$ an ϵ such that $|p| > R + \epsilon$, the power series diverges for all p with $|p| > R$. In particular, in case $R = 0$ we conclude as before that the series diverges for all $|p| \neq 0$. And finally for $R = \infty$ we have $\lim_{n\to\infty} |a_n|^{1/n} = 0$; it follows that for all $\epsilon > 0$ and sufficiently large n, $|a_n p^n| \leq (\epsilon|p|)^n$. On the right-hand side we have for $\epsilon|p| < 1$ a comparison series that converges for all $|p|$ with sufficiently small ϵ.

Example. For which values of p does the quaterion power series $\sum_{n=2}^{\infty}(-1)^n \frac{(p-i+k)^n}{4^n \ln(n)}$ converge? What is the radius of convergence?

Solution. By the ratio test (4.6), we have

$$\lim_{n\to\infty} \sup \frac{\left|\frac{(-1)^{n+1}}{4^{n+1}\ln(n+1)}(p - i + k)^{n+1}\right|}{\left|\frac{(-1)^n}{4^n \ln(n)}(p - i + k)^n\right|} = \frac{1}{4}\lim_{n\to\infty}\sup \frac{\ln(n)}{\ln(n+1)}|p - i + k|$$

$$= \frac{1}{4}|p - i + k|.$$

Thus the series converges absolutely for $|p - i + k| < 4$. The ball of convergence is $|p - i + k| < 4$, and the radius of convergence is $R = 4$. We highlight the fact that, at $p - i + k = -4$, the series

$$\sum_{n=2}^{\infty}(-1)^n \frac{(-4)^n}{4^n \ln(n)} = \sum_{n=2}^{\infty} \frac{1}{\ln(n)}$$

diverges because for $n \geq 2$ one has that $\frac{1}{n} \leq \frac{1}{\ln(n)}$. For $p - i + k = 4$, $\sum_{n=2}^{\infty}(-1)^n \frac{1}{\ln(n)}$ is an alternating series and hence converges. Therefore, the series converges at all points p such that $|p - i + k| = 4$ except at $p = -4 + i - k$.

Exercise 4.120. *Do the quaternion power series $\sum_{n=0}^{\infty} a_n p^n$, $\sum_{n=1}^{\infty} n a_n p^{n-1}$, and $\sum_{n=0}^{\infty} \frac{a_n}{n+1} p^{n+1}$ have the same radius of convergence? Justify your answer.*

Solution. They have the same radius of convergence since $\lim n^{\frac{1}{n}} = \lim(n + 1)^{\frac{1}{n}} = 1$.

Exercise 4.121. *Let $\sum_{n=0}^{\infty} a_n p^n$ and $\sum_{n=0}^{\infty} b_n p^n$ be power series with radii of convergence R_1 and R_2, respectively. Assume that there is a positive constant M such that $|a_n| \leq M |b_n|$ for all but finitely many n. Prove that $R_1 \geq R_2$.*

4.15   Theoretical Radius

For a quaternion power series $\sum_{n=0}^{\infty} a_n(p-q)^n$, the limit (4.6) depends on the quaternion coefficients a_n only. Hence, if

(i) $\lim\limits_{n\to\infty} \dfrac{|a_{n+1}|}{|a_n|} = L_2 \neq 0$, the radius of convergence is $R = \frac{1}{L_2}$;

(ii) $\lim\limits_{n\to\infty} \dfrac{|a_{n+1}|}{|a_n|} = 0$, $R = \infty$;

(iii) $\lim\limits_{n\to\infty} \dfrac{|a_{n+1}|}{|a_n|} = \infty$, $R = 0$.

In case the ratio test fails, or becomes difficult to apply, we can often use the root test (4.5), which says that $R = \dfrac{1}{\lim_{n\to\infty} \sup \sqrt[n]{|a_n|}}$, provided this limit exists.

Exercise 4.122. *Show that if $\{a_n\}$ is a sequence of nonzero quaternion numbers such that $\lim_{n\to\infty} \sup \frac{|a_{n+1}|}{|a_n|} = L$, then $\lim_{n\to\infty} \sup \sqrt[n]{|a_n|} = L$.*

4.16   The Arithmetic of Quaternion Power Series

We now define certain arithmetic operations on one or more quaternion power series:

(i) A quaternion power series $\sum_{n=0}^{\infty} a_n(p-q)^n$ can be multiplied by a nonzero quaternion constant λ without affecting its convergence or divergence;

(ii) A quaternion power series $\sum_{n=0}^{\infty} a_n(p-q)^n$ converges absolutely within its ball of convergence. As a consequence, within the ball of convergence the terms of the series can be rearranged and the rearranged series has the same sum S as the original series;

(iii) Two quaternion power series $\sum_{n=0}^{\infty} a_n(p-q)^n$ and $\sum_{n=0}^{\infty} b_n(p-q)^n$ can be added and subtracted by adding or subtracting like terms, namely:

$$\sum_{n=0}^{\infty} a_n(p-q)^n \pm \sum_{n=0}^{\infty} b_n(p-q)^n = \sum_{n=0}^{\infty} (a_n \pm b_n)(p-q)^n.$$

If both series have the same nonzero radius of convergence R, the radius of convergence of $\sum_{n=0}^{\infty} (a_n \pm b_n)(p-q)^n$ is R. If one series has radius of convergence $R_1 > 0$ and the another one has radius of convergence $R_2 > 0$, where $R_1 \neq R_2$, then the radius of convergence of $\sum_{n=0}^{\infty} (a_n \pm b_n)(p-q)^n$ is the smaller of the two numbers R_1 and R_2.

Exercise 4.123. *Let a_n be a quaternion sequence such that $|a_n| \geq 2^n$ for all n. What can you say about the radius of convergence of $\sum_{n=0}^{\infty} a_n p^n$?*

Solution. Since $\lim_{n\to\infty} \sup |a_n|^{\frac{1}{n}} \geq 2$, one has $R \in \left]0, \frac{1}{2}\right]$.

Exercise 4.124. *Suppose the radius of convergence of $\sum_{n=0}^{\infty} a_n p^n$ is R. What is the radius of convergence of each of the following quaternion power series?* (a) $\sum_{n=0}^{\infty} n^2 a_n p^n$;   (b) $\sum_{n=0}^{\infty} a_n p^{2n}$;   (c) $\sum_{n=0}^{\infty} a_n^2 p^n$.

Solution. (a) $R = \dfrac{1}{\lim_{n\to\infty} \sup |a_n|^{\frac{1}{n}}}$;   (b) $R = \dfrac{1}{\lim_{n\to\infty} \sup |a_n|^{\frac{1}{2n}}}$;   (c) $R = \dfrac{1}{\lim_{n\to\infty} \sup |a_n|^{\frac{2}{n}}}$.

Next we consider infinite products of sequences of quaternion numbers.

4.17   Infinite Products of Quaternion Numbers

If $\{p_\nu\}$ is a sequence of quaternion numbers, the sequence

$$\left\{ \prod_{\nu=k}^{n} p_\nu \right\}_{\mathcal{R}} = p_k\, p_{k+1} \cdots p_n$$

of products is called a(n) (infinite) quaternion right product with the factors p_ν. In an analogous way, the sequence

$$\left\{ \prod_{\nu=k}^{n} p_\nu \right\}_{\mathcal{L}} = p_n\, p_{n-1} \cdots p_k$$

of products is called a(n) (infinite) quaternion left product with the factors p_ν. For the reader's convenience, from now on we only deal with the above quaternion right product, which for simplicity we call quaternion product. It will be simply denoted by $\{\prod_{\nu=k}^{n} p_\nu\}$. We write $\prod_{\nu=k}^{\infty} p_\nu$; usually, $k = 0$ or $k = 1$. By analogy with the series, the product is said to converge when the limit of the sequence of partial products exists (as a finite limit in $\mathbb{H}$) and is nonzero. Otherwise the product is said to diverge.

4.18   Absolute and Conditional Convergence of a Quaternion Product

We say that $\prod_{\nu=k}^{\infty} p_\nu$ converges absolutely if $\prod_{\nu=k}^{\infty} |p_\nu|$ converges. If $\prod_{\nu=k}^{\infty} p_\nu$ converges absolutely, then $\prod_{\nu=k}^{\infty} p_\nu$ converges, but the converse is false. If $\prod_{\nu=k}^{\infty} p_\nu$ converges but $\prod_{\nu=k}^{\infty} |p_\nu|$ does not, then we say that $\prod_{\nu=k}^{\infty} p_\nu$ converges conditionally.

Example. Investigate the convergence of the quaternion product $\prod_{n=1}^{\infty} p_n$, where $p_n = e^{\frac{1}{n^2}}(1 + i + j + k)$.

Solution. A straightforward computation shows that

$$\prod_{n=1}^{\infty} |p_n| \le 2 \prod_{n=1}^{\infty} e^{\frac{1}{n^2}} < \infty.$$

Then the product converges, and so it converges absolutely.

Exercise 4.125. *Let $\prod_{n=1}^{\infty} p_n$ be a convergent quaternion product. Prove that $\prod_{n=1}^{\infty} e^{\frac{1}{n^3}} p_n$ is convergent.*

Exercise 4.126. *Let p be a nonnull quaternion. Show that the quaternion product $\prod_{n=1}^{\infty} \left\{ 1 - \left(1 - \frac{1}{n}\right)^{-n} p^{-n} \right\}$ converges for all points p situated outside the unit ball centered at the origin.*

4.19 Advanced Practical Exercises

1. Let $\{p_n\}$ be a sequence of real quaternions such that $\lim_{n\to\infty} p_n = p$, with $p \in \mathbb{H}$. Show that $\lim_{n\to\infty} 2^{-n} \sum_{m=1}^{n+1} \binom{n}{m} p_m = p$.
2. Prove that the series

$$\sum_{n=1}^{\infty} p_n, \qquad p_n = a_n + b_n i + c_n j + d_n k, \qquad a_n, b_n, c_n, d_n \in \mathbb{R},$$

   is convergent if and only if the series $\sum_{n=1}^{\infty} a_n$, $\sum_{n=1}^{\infty} b_n$, $\sum_{n=1}^{\infty} c_n$, and $\sum_{n=1}^{\infty} d_n$ are convergent.
3. Let $\sum_{n=1}^{\infty} p_n$ and $\sum_{n=1}^{\infty} q_n$ be convergent quaternion series. Let also $a, b \in \mathbb{H}$. Prove that the quaternion series $\sum_{n=1}^{\infty} (ap_n + q_n b)$ is convergent.
4. Investigate the convergence and find the sum of the following series:

   (a) $\displaystyle\sum_{n=1}^{\infty} \left(\frac{1}{(2n-1)(2n+5)} + \frac{1}{n(n+1)(n+2)} i + \arctan \frac{1}{2n} j + \frac{1}{2n^2 - 1} k \right);$

   (b) $\displaystyle\sum_{n=1}^{\infty} \left(\frac{2n+1}{n^2(n+1)^2} + \frac{1}{n(n+1)} i - \frac{1}{(n+1)(n+2)} j + \frac{1}{(n+3)(n+4)} k \right).$

   Hint. Use the previous exercise.

5. Let p be a real quaternion and m a positive integer. Expand $(1_{\mathbb{H}} - p)^{-m}$ in powers of p.
6. Investigate the absolute convergence of the following series:

   (a) $\displaystyle\sum_{n=1}^{\infty} \left(\frac{1}{n^2} + \frac{1}{n^3} i + (\sqrt{n + \arctan n^\alpha} - \sqrt{n}) j + \left(\left(n \arctan \frac{1}{n} \right)^{n^\alpha} - 1 \right) k \right);$

(b) $\displaystyle\sum_{n=1}^{\infty}\left(\left|\ln\cos\left(\tan\frac{1}{n}-\arctan\frac{1}{n}\right)\right|^{\alpha}+\frac{1}{n^3}j+\frac{\left(e-\left(1+\frac{1}{n}\right)^n\right)^{\alpha}}{(1-\cos\frac{1}{n})^2}\left(1+\tan\frac{1}{n}\right)k\right);$

(c) $\displaystyle\sum_{n=1}^{\infty}\left(\frac{1}{n^4}i+\frac{1}{n^2}j+\left(e^{n\sin\frac{1}{n^3}}-n^{\alpha}-1\right)k\right);$

(d) $\displaystyle\sum_{n=1}^{\infty}\left(\frac{1}{n^4}i+\left(\frac{e^3-\left(1+\frac{3}{n}\right)^n}{n}\right)^{\alpha}j+\left(\frac{\pi}{2n+2}-\cos\frac{\pi n}{2n+2}\right)^{\alpha}\ln\cos\frac{1}{n}k\right);$

(e) $\displaystyle\sum_{n=1}^{\infty}\left(\frac{1}{n^2}+\left(e^{n\arctan\frac{1}{n^2}}-1-\tan\frac{1}{n}-n^{\alpha}\right)i-\frac{1}{n^3}j+\left(\left(e^{\frac{1}{n}}-\sinh\frac{1}{n}\right)^{n^{\alpha}}-1\right)k\right);$

(f) $\displaystyle\sum_{n=1}^{\infty}\left(\frac{1}{n^2}i+\left(\ln\left(1+\tan^2\frac{1}{n^{\alpha}}\right)-\ln\cos\frac{1}{n^{\beta}}\right)j+\right.$
$\left.\left(\cos\left(\cos\frac{1}{n}\right)-\cos\left(\cosh\frac{1}{n}\right)\right)^{\alpha}\left(\ln\frac{\arctan\frac{1}{n}}{\tan\frac{1}{n}}\right)^{\beta}k\right).$

7. Investigate the convergence of the following series:

(a) $\displaystyle\sum_{n=1}^{\infty}\left(\frac{1}{n^2+2n+2}+\left(\frac{n+1}{n^2+1}\right)^2i+\frac{1}{n}\left(\frac{3}{4}\right)^nj+\frac{\left(\frac{n+1}{n}\right)^{n^2}}{3^n}k\right);$

(b) $\displaystyle\sum_{n=1}^{\infty}\left(\frac{1}{n(n-1)}+\left(\frac{n}{2n+2}\right)^ni+\ln\left(1+\frac{1}{n^2}\right)j+\frac{3^{n-1}}{n^2}k\right);$

(c) $\displaystyle\sum_{n=1}^{\infty}\left(\ln\left(1+\frac{1}{n^2}\right)+\frac{3^{n-1}}{n^n}i+\frac{(n!)^2}{(2n)!}j-\left(\frac{n}{3(n+1)}\right)^nk\right);$

(d) $\displaystyle\sum_{n=1}^{\infty}\left(\frac{n}{1+n^2}+\frac{1}{n}\left(\frac{1}{2}\right)^ni+\frac{1}{\sqrt{n+1}}j+\frac{n+1}{n^2+4}k\right);$

(e) $\displaystyle\sum_{n=1}^{\infty}\left(\left(\frac{n}{n+1}\right)^n+\ln\left(1+\frac{2}{n^2}\right)i+\frac{4^nn!}{n^n}j+(-1)^n\frac{1}{\ln(n+1)}k\right);$

(f) $\displaystyle\sum_{n=1}^{\infty}\left(\left(\frac{5n}{n+2}\right)^ni-\frac{4^{n-1}}{n^n}k\right).$

8. Investigate the convergence of the following infinite products:

(a) $\displaystyle\prod_{n=1}^{\infty}\left(\frac{1}{n(n+5)}+i+\frac{1}{n^2}j+\frac{1}{n^3}k\right);$

(b) $\displaystyle\prod_{n=3}^{\infty}\left(1+\frac{1}{n^2-4}i+\frac{1}{n^4}j+\frac{1}{n^2}k\right);$

(c) $\displaystyle\prod_{n=1}^{\infty}\left(\frac{3n^2+3n+1}{n^3(n+1)^3}+i+j+k\right);$

(d) $\displaystyle\prod_{n=1}^{\infty}\left(1+\frac{1}{n^3}i+\frac{1}{n^4}j+\frac{1}{2^n\cotan\frac{x}{2^n}}\right),\quad 0<x<\frac{\pi}{2};$

(e) $\displaystyle\prod_{n=1}^{\infty}\left(1 + \arctan\frac{x^{3n}+1}{x^{2n}+1}i + \frac{1}{n^2}j + \frac{1}{n^3}k\right);$

(f) $\displaystyle\prod_{n=1}^{\infty}\left(1 + \sqrt{n}\,i + \left(\frac{n-1}{n+1}\right)^{n(n-1)}j + k\right);$

(g) $\displaystyle\prod_{n=1}^{\infty}\left(1 + \frac{\left(5+(-1)^n\right)^n}{n^2 7^n}i + \left(\frac{n^2+3}{n^2+4}\right)^{n^3+1}j + \frac{n^5}{5^{\sqrt{n}}}k\right);$

(h) $\displaystyle\prod_{n=1}^{\infty}\left(\frac{1}{\ln n} + i + \frac{n^2}{2^{n^2}}j + \frac{1}{\ln n^{\ln n}}k\right);$

(i) $\displaystyle\prod_{n=1}^{\infty}\left(1 + \frac{(3n)^{n+1}}{(n+1)^{3n}}i + \frac{1}{\sqrt{n}}\sin\frac{1}{n}j + \tan\frac{1}{\sqrt{n}}\frac{1}{n^{\alpha}}k\right), \quad \alpha > 0;$

(j) $\displaystyle\prod_{n=1}^{\infty}\left(\frac{n^4+4n^2+1}{2^n} + \frac{1}{3^n-n^2}i + \frac{(n!)^3}{(3n)!}j + k\right);$

(k) $\displaystyle\prod_{n=1}^{\infty}\left(\left(\frac{n+2}{2n+1}\right)^{n^2} + \frac{1}{\sqrt{n}}\ln^l\left(\frac{n+1}{n-1}\right)i + \left(\cos\frac{n^2+1}{n^2-1}\right)^n j + k\right), \quad l > 0;$

(l) $\displaystyle\prod_{n=1}^{\infty}\left(1 + \frac{n^3+1}{\left(1+\frac{1}{n}\right)^n}i + \arccos\frac{1}{n^2+4}j + \arccos\frac{n^3}{n^2+4}k\right);$

(m) $\displaystyle\prod_{n=1}^{\infty}\left(\operatorname{arccotan}\frac{n^3-1}{n+4} + \arctan\frac{n^3-1}{n+4}i + j + 2^n\left(\frac{n+2}{n+3}\right)^n k\right);$

(n) $\displaystyle\prod_{n=1}^{\infty}\left(\left(\arcsin\frac{1}{3^n}\right)\frac{n!}{(2n)!!} + \frac{\arcsin\frac{n-1}{n}}{n\sqrt{\ln(n+2)}}i + \left(1-\frac{\ln n}{n}\right)^n j + k\right);$

(o) $\displaystyle\prod_{n=1}^{\infty}\left(1 + n\left(2^{\frac{1}{n}}-1\right)k\right).$

Exponents and Logarithms **5**

The real exponential and logarithmic functions play an important role in advanced mathematics, including applications to calculus, differential equations, and complex analysis. In this chapter we use the properties of quaternions described in the previous chapters to define and study the quaternionic analogues of these functions. It turns out that exponential and logarithmic quaternion functions can be defined because quaternions form a division algebra. To make sense of this, it will be convenient to define some of their more common properties, which as in the real and complex cases, should be familiar to the reader.

New aspects come into view when we consider quaternion-valued functions depending on a quaternion variable, that is, when we study functions whose domain and range are subsets of the four-dimensional quaternion space. As a first example we introduce the quaternion exponential function e^p for a given real quaternion p, which has many similarities with the real and complex exponential functions. We then proceed by introducing the quaternion logarithm $\ln(q)$, which will be used to solve exponential equations of the form $e^p = q$ for a quaternion q. It will be shown that when q is a fixed nonzero quaternion number there are infinitely many solutions to the equation $e^p = q$. The principal value of the quaternion logarithm will be defined to be a single-valued quaternion function whose argument lies in the interval $[0, \pi]$. This principal value quaternion function will be shown to be an inverse function of the quaternion exponential function defined on a suitably restricted domain of the quaternion space. We will illustrate this point with the aid of some examples and exercises, since our experience has shown that this approach is useful for someone who is introduced to the subject for the first time. Further examples may be found in the exercises at the end of the chapter.

At this stage it is convenient to define some of the more common functions of a quaternion variable. This, as in the real and complex cases, will be familiar to the reader.

J.P. Morais et al., *Real Quaternionic Calculus Handbook*,
DOI 10.1007/978-3-0348-0622-0_5, © Springer Basel 2014

5.1   Quaternion Natural Exponential Function

The function e^p defined by

$$e^p := e^{p_0}\Big(\cos |\mathbf{p}| + \mathrm{sgn}(\mathbf{p}) \sin |\mathbf{p}|\Big) \tag{5.1}$$

is called the quaternion natural exponential function, and is defined to be merely a notational convenience given its similarity with the de Moivre's formula.[1] The number "e" is a mathematical constant, the base of the natural logarithm, and is known as the Euler number.[2] When p is a real or complex number, the definition of e^p does naturally yields the usual exponential function of real and complex numbers. The previous representation is explained in the following calculation. For this purpose it will be necessary to define $e^p := \sum_{l=0}^{\infty} \frac{p^l}{l!}$. The reader should notice that this series converges for all p, in analogy to the complex case, since we have $|p^l| \leq |p|^l$ for any quaternion p. Since $e^{|p|}$ converges, the comparison test yields that e^p converges for all p. Clearly, the series expansions

$$e^{p_0} = \sum_{l=0}^{\infty} \frac{p_0^l}{l!} \quad \text{and} \quad e^{\mathbf{p}} = \sum_{l=0}^{\infty} \frac{\mathbf{p}^l}{l!}$$

converge. It is easy to show that for an arbitrary $\epsilon > 0$ it is always possible to find a sufficiently large number N such that, for any $r, s > N$,

$$\left| \sum_{l=r}^{s} \frac{\mathbf{p}^l}{l!} \right| \leq \sum_{l=r}^{s} \frac{|\mathbf{p}|^l}{l!} < \epsilon.$$

Therefore, taking the Cauchy product of e^{p_0} and $e^{\mathbf{p}}$ one obtains

$$\sum_{l=0}^{\infty} \left\{ \sum_{n=0}^{l} \frac{p_0^n}{n!} \frac{\mathbf{p}^{l-n}}{(l-n)!} \right\} = \sum_{l=0}^{\infty} \frac{1}{l!} \sum_{n=0}^{l} \binom{l}{n} p_0^n \mathbf{p}^{l-n}$$

$$= \sum_{l=0}^{\infty} \frac{(p_0 + \mathbf{p})^l}{l!}.$$

Consequently, $e^{p_0} e^{\mathbf{p}} = e^{p_0 + \mathbf{p}} := e^p$. For the remaining term $e^{\mathbf{p}}$, one has, explicitly,

---

[1] It is not trivial to define the quaternion exponential function. Here, we prefer to do this through a power series expansion.

[2] This constant is defined as the limit $e = \lim\limits_{n \to \infty} \left(1 + \dfrac{1}{n}\right)^n$.

$$\sum_{l=0}^{\infty} \frac{\mathbf{p}^l}{l!} = \sum_{n=0}^{\infty} \frac{\mathbf{p}^{2n}}{(2n)!} + \sum_{n=0}^{\infty} \frac{\mathbf{p}^{2n+1}}{(2n+1)!}$$

$$= \sum_{n=0}^{\infty} (-1)^n \frac{|\mathbf{p}|^{2n}}{(2n)!} + \frac{\mathbf{p}}{|\mathbf{p}|} \sum_{n=0}^{\infty} (-1)^{n+1} \frac{|\mathbf{p}|^{2n+1}}{(2n+1)!}$$

$$= \cos|\mathbf{p}| + \mathrm{sgn}(\mathbf{p}) \sin|\mathbf{p}|.$$

From the practical point of view, the quaternion exponential function is an example of one that is defined by specifying its scalar and vectors parts. More precisely, the scalar and vector parts of e^p are, respectively, $\mathrm{Sc}(e^p) = e^{p_0} \cos|\mathbf{p}|$ and $\mathrm{Vec}(e^p) = e^{p_0} \mathrm{sgn}(\mathbf{p}) \sin|\mathbf{p}|$. Thus, the values of this quaternion function are found by expressing the point p as $p = t + xi + yj + zk$, and then substituting the values of t, x, y and z in the given expression. We record now some useful properties of the quaternion exponential function.

(i) $e^p \neq 0_{\mathbb{H}}$, for all $p \in \mathbb{H}$;

(ii) $e^{-p} e^p = 1_{\mathbb{H}}$, $e^{\mathrm{sgn}(\mathbf{p})\pi} = -1_{\mathbb{H}}$;

(iii) $(e^p)^n = e^{np}$ for $n = 0, \pm 1, \pm 2, \ldots$ (de Moivre's formula);

(iv) In general $e^p e^q \neq e^{p+q}$, unless p and q commute.

For the proof we consider two arbitrary quaternions p and q. Then $\sum_{l=0}^{\infty} \frac{p^l}{l!}$ and $\sum_{l=0}^{\infty} \frac{q^l}{l!}$ converge. When p and q commute, the Cauchy product of the series expansions of e^p and e^q yields

$$\sum_{l=0}^{\infty} \frac{(p+q)^l}{l!} = \sum_{l=0}^{\infty} \frac{1}{l!} \sum_{n=0}^{l} \binom{l}{j} p^n q^{l-n}$$

$$= \sum_{l=0}^{\infty} \left\{ \sum_{n=0}^{l} \frac{p^n}{n!} \frac{q^{l-n}}{(l-n)!} \right\}$$

$$= \sum_{n=0}^{\infty} \frac{p^n}{n!} \sum_{m=0}^{\infty} \frac{q^m}{m!} = e^p e^q.$$

In particular, $e^p e^{-p} = e^{0_{\mathbb{H}}} = 1_{\mathbb{H}}$, so that $e^p \neq 0_{\mathbb{H}}$ for all $p \in \mathbb{H}$. Indeed, by induction, $(e^p)^n = e^{np}$, where n is any positive or negative integer. For the last statement, take for example $e^{\pi i} e^{\pi j} = (-1)(-1) = 1_{\mathbb{H}}$, and $e^{\pi i + \pi j} = \cos(\pi \sqrt{2}) + \frac{i+j}{\sqrt{2}} \sin(\pi \sqrt{2}) \neq 1_{\mathbb{H}}$.

Exercise 5.127. *Find the values of the quaternion exponential function e^p at the following points:* (a) $p = 1_{\mathbb{H}}$; (b) $p = i + j + k$; (c) $p = (1 - i + j)^5$; (d) $p = (1 - i + k)^n (-\frac{1}{3} + \frac{1}{3}i - \frac{1}{3}j)^n$.

Solution. (a) e;   (b) $\cos\sqrt{3} + \frac{i+j+k}{\sqrt{3}} \sin\sqrt{3}$;   (c) $\cos(11\sqrt{2}) + \frac{i-j}{\sqrt{2}} \sin(11\sqrt{2})$;
(d) if $n = 2l$ then $e^p = e^{(-1)^l}$, and if $n = 2l + 1$, then $e^p = \cos 1 + i(-1)^l \sin 1$.

Exercise 5.128. *Compute* $\sum_{l=0}^{\infty} \frac{(i-j+k)^l}{l!}$.

Solution. e^{i-j+k}.

Exercise 5.129. *Is the quaternion exponential function periodic?*

Example. Show that the usual limit representation is valid for the quaternion natural exponential function: $e^p = \lim_{n\to\infty} \left(1 + \frac{p}{n}\right)^n$, for any quaternion p.

Solution. Let $p = a + bi + cj + dk$, where $a, b, c, d \in \mathbb{R}$. Hence

$$1 + \frac{p}{n} = 1 + \frac{a}{n} + \frac{b}{n}i + \frac{c}{n}j + \frac{d}{n}k.$$

Making use of the polar form of a quaternion (1.4) we obtain

$$\left(1 + \frac{p}{n}\right)^n$$

$$= \left(1 + \frac{2a}{n} + \frac{a^2 + b^2 + c^2 + d^2}{n^2}\right)^{\frac{n}{2}} \left[\cos\left(n \arccos \frac{1 + \frac{a}{n}}{\sqrt{1 + \frac{2a}{n} + \frac{a^2+b^2+c^2+d^2}{n^2}}}\right) \right.$$

$$\left. + \frac{bi + cj + dk}{\sqrt{b^2 + c^2 + d^2}} \sin\left(n \arccos \frac{1 + \frac{a}{n}}{\sqrt{1 + \frac{2a}{n} + \frac{a^2+b^2+c^2+d^2}{n^2}}}\right) \right].$$

Then

$$\lim_{n\to\infty} \left(1 + \frac{p}{n}\right)^n = \lim_{n\to\infty} \left(1 + \frac{2a}{n} + \frac{a^2 + b^2 + c^2 + d^2}{n^2}\right)^{\frac{n}{2}} \times$$

$$\times \left[\cos\left(n \arccos \frac{1 + \frac{a}{n}}{\sqrt{1 + \frac{2a}{n} + \frac{a^2+b^2+c^2+d^2}{n^2}}}\right) \right.$$

$$\left. + \frac{bi + cj + dk}{\sqrt{b^2 + c^2 + d^2}} \sin\left(n \arccos \frac{1 + \frac{a}{n}}{\sqrt{1 + \frac{2a}{n} + \frac{a^2+b^2+c^2+d^2}{n^2}}}\right) \right].$$

Straightforward computations show that

$$\lim_{n\to\infty} n \arccos \frac{1+\frac{a}{n}}{\sqrt{1+\frac{2a}{n}+\frac{a^2+b^2+c^2+d^2}{n^2}}} = \sqrt{b^2+c^2+d^2}$$

and

$$\lim_{n\to\infty}\left(1+\frac{2a}{n}+\frac{a^2+b^2+c^2+d^2}{n^2}\right)^{\frac{n}{2}} = e^a.$$

Consequently

$$\lim_{n\to\infty}\left(1+\frac{p}{n}\right)^n = e^a\left(\cos\sqrt{b^2+c^2+d^2}+\frac{bi+cj+dk}{\sqrt{b^2+c^2+d^2}}\sin\sqrt{b^2+c^2+d^2}\right)$$

$$= e^p.$$

Exercise 5.130. *Compute* $e^{p_+}+(p_-,q_+)+[p,q]-2q_-(3p_++2q_+)$, *where* $p = 1+2i-j-k$ *and* $q = 1-i-j+k$.

Solution. $p_+ = \frac{3}{2}i-\frac{3}{2}j$, $p_- = 1+\frac{1}{2}i+\frac{1}{2}j-k$, $q_+ = 1+k$, $q_- = -i-j$, $e^{p_+} = \cos(\frac{3\sqrt{2}}{2})+(\frac{\sqrt{2}}{2}i-\frac{\sqrt{2}}{2}j)\sin(\frac{3\sqrt{2}}{2})$, $(p_-,q_+) = -i+2k$, $[p,q] = -1+3i-j+3k$, $2q_-(3p_++2q_+) = -8i+18k$, $-1+\cos(\frac{3\sqrt{2}}{2})+(\frac{\sqrt{2}}{2}\sin(\frac{3\sqrt{2}}{2})+10)i-(\frac{\sqrt{2}}{2}\sin(\frac{3\sqrt{2}}{2})+1)j-13k$.

Exercise 5.131. *Compute the quaternion limit* $\lim_{p\to 0_{\mathbb{H}}}\frac{p}{e^p-1_{\mathbb{H}}}$.

Solution. $1_{\mathbb{H}}$.

5.2 Modulus, Argument, and Conjugate of the Quaternion Exponential Function

The modulus, argument, and conjugate of the quaternion exponential function are easily determined from (5.1). To do this, we start by expressing the quaternion $w = e^p$ in polar form, with a quaternion modulus and an argument: $w = |w|\left(\cos\theta + \text{sgn}(\mathbf{w})\sin\theta\right)$. Now one can easily see that $|w| = e^{p_0}$ and $\theta = |\mathbf{p}| + 2\pi n$, for $n = 0,\pm 1,\pm 2,\ldots$. Since $|w| = |e^p|$ and θ is an argument of w, we have: $|e^p| = e^{p_0}$ and $\arg(e^p) = |\mathbf{p}| + 2\pi n$, $n = 0,\pm 1,\pm 2,\ldots$. For the conjugate of the quaternion function e^p a formula is found using properties of the real cosine and sine functions:

$$\overline{e^p} = e^{p_0}\left(\cos|\mathbf{p}| - \text{sgn}(\mathbf{p})\sin|\mathbf{p}|\right)$$

$$= e^{p_0}\left(\cos|-\mathbf{p}| - \text{sgn}(\mathbf{p})\sin|-\mathbf{p}|\right)$$

$$= e^{Sc(\overline{p})}\left(\cos|\overline{\mathbf{p}}| + \text{sgn}(\overline{\mathbf{p}})\sin|\overline{\mathbf{p}}|\right)$$

$$= e^{\overline{p}}.$$

Exercise 5.132. *Write the given expression in terms of* t, x, y *and* z: (a) $(i + j + k)e^p$;   (b) e^{p^2-p};   (c) $\arg(p + \overline{p})$;   (d) e^{p^2};   (e) $e^{(p^{-1}+p)^2}$.

Exercise 5.133. *Evaluate* $|e^p|$ *and show that* $|e^p| \le e^{|p|}$ *for any real quaternion* p. *In addition, verify the validity of the following inequalities:* (a) *For all* $p \in \mathbb{H}$, *we have* $|e^p - 1_\mathbb{H}| \le e^{|p|} - 1 \le |p|e^{|p|}$;   (b) *For all* $p \in \mathbb{H}$ *with* $|p| \le 1$, *we have* $|e^p - 1_\mathbb{H}| \le 2|p|$.

In the following we introduce the quaternion natural logarithm function $\ln(p)$, which is motivated by the need to define the "inverse" of the quaternion natural exponential function e^p. More precisely, we define $\ln(p)$ to be any quaternion number such that $e^{\ln(p)} = p$ and $\ln(e^p) = p$. This is too much to hope for. We shall discuss this matter in further detail in the remainder of the section.

5.3   Quaternion Natural Logarithm Function

The quaternion natural logarithm function $\ln(p)$ is given by

$$\ln(p) := \log_e|p| + \text{sgn}(\mathbf{p})\arg(p). \tag{5.2}$$

Here $\log_e|p|$ is the usual real natural logarithm of the positive number $|p|$ (and hence is defined unambiguously). This quaternion function is another example of one that is defined by specifying its scalar and vectors parts. More precisely, the scalar and vector parts of $\ln(p)$ are, respectively, $\text{Sc}(\ln(p)) = \log_e|p|$ and $\text{Vec}(\ln(p)) = \text{sgn}(\mathbf{p})\arg(p)$. Because there are intrinsically infinitely many arguments of p, it is clear that representation (5.2) gives infinitely many solutions w to the equation $e^w = p$ whenever p is a nonzero quaternion number. By switching to polar form in (5.2), we obtain the following alternative description of the quaternion logarithm:

$$\ln(p) = \begin{cases} \log_e|p| + \text{sgn}(\mathbf{p})\left(\arccos\frac{p_0}{|p|} + 2\pi n\right), & \text{if } |\mathbf{p}| \ne 0, \\ \log_e|p_0|, & \text{if } |\mathbf{p}| = 0, \end{cases}$$

$$= \begin{cases} \log_e|p| + \text{sgn}(\mathbf{p})\left(\arctan\frac{|\mathbf{p}|}{p_0} + 2\pi n\right), & \text{if } p_0 > 0, \\ \log_e|\mathbf{p}| + \text{sgn}(\mathbf{p})\left(\frac{\pi}{2} + 2\pi n\right), & \text{if } p_0 = 0, \end{cases}$$

where $n = 0, \pm1, \pm2, \dots$. Observe that the different values of $\ln(p)$ all have the same scalar part and that their vector parts differ by $2\pi n$. Each value of n determines what is known as a branch (or sheet), a single-valued component of the multiple-valued logarithmic quaternion function. When $n = 0$, we have a special situation.

Example. Find all quaternion solutions p to the equation $e^p = i$.

Solution. For each equation $e^p = q$, the set of solutions is given by $p = \ln(q)$. Setting $q = i$ we deduce that $|q| = 1$ and $\arg(q) = \frac{\pi}{2} + 2\pi n$ for $n = 0, \pm 1, \pm 2, \ldots$. Thus, from representation (5.2) we establish:

$$p = \ln(i) = \frac{(4n + 1)\pi}{2} i, \quad n = 0, \pm 1, \pm 2, \ldots.$$

Therefore, each of the values: $p = \ldots, -\frac{7\pi}{2}i, -\frac{3\pi}{2}i, \frac{\pi}{2}i, \frac{5\pi}{2}i, \ldots$ satisfies the equation $e^p = i$.

Example. Let p be a real quaternion such that $|\mathbf{p}| \neq 0$. The following relation holds:

$$\ln(p) = \log_e |p| + \operatorname{sgn}(\mathbf{p})\left(\arccos \frac{p_0}{|p|} + 2\pi n\right),$$

$$= \log_e |p| + \operatorname{sgn}(\mathbf{p})\left(\frac{\pi}{2} - \arcsin \frac{p_0}{|p|} + 2\pi n\right),$$

$$= \log_e |p| + \operatorname{sgn}(\mathbf{p})\left(-\arcsin \frac{p_0}{|p|} + \frac{(4n + 1)}{2}\pi\right), \quad n \in \mathbb{Z}.$$

Exercise 5.134. *Let $p = 1 - i + j - k$. Find $\ln(p)$.*

Solution. $\log_e 2 + \frac{i - j + k}{\sqrt{3}}\left(\frac{\pi}{6} - \frac{4n+1}{2}\pi\right)$.

Exercise 5.135. *Find two quaternions p and q such that $\ln(pq) \neq \ln(p) + \ln(q)$. In addition, find p and q such that $\ln(pq) = \ln(p) + \ln(q)$.*

Exercise 5.136. *Find all quaternion values of p satisfying the given equation:*
(a) $e^p = i$;   (b) $e^p = 1 + i + j + k$;   (c) $e^{(p^{-1})} = 1$;   (d) $e^{p + \bar{p}} = 1$;
(e) $e^p + e^{-p} = (-i + j)^3$.

Exercise 5.137. *Does the relation $\ln(p^n) = n \ln(p)$ hold for all nonzero quaternion numbers p and all integers n? Justify your answer.*

5.4   Limits of the Quaternion Exponential and Logarithm Functions

Let $\{p_n\}$ be a sequence of elements of $\mathbb{H}$ and $n \in \mathbb{N}$. The following statements are valid:

(i) If $\lim_{n\to\infty} p_n = \infty$, then $\lim_{n\to\infty} e^{p_n} = \infty$;

(ii) If $\lim_{n\to\infty} p_n = \infty$, then $\lim_{n\to\infty} e^{\frac{1}{p_n}} = 1_{\mathbb{H}}$;

(iii) If $\lim_{n\to\infty} p_n = \infty$, then $\lim_{n\to\infty} (\pm p_n e^{-p_n}) = 0_{\mathbb{H}}$;

(iv) If $\lim_{n\to\infty} p_n = \infty$, then $\lim_{n\to\infty} \ln(p_n) = \infty$ and $\lim_{n\to\infty} \dfrac{\ln(p_n)}{p_n} = 0_{\mathbb{H}}$;

(v) If $\lim_{n\to\infty} p_n = 0_{\mathbb{H}}$, then $\lim_{n\to\infty} \ln(p_n) = -\infty$;

(vi) If $\lim_{n\to\infty} p_n = p$, then $\lim_{n\to\infty} \ln(p_n) = \ln(p)$;

(vii) If $\lim_{n\to\infty} p_n = \infty$, then $\lim_{n\to\infty} p_n^{-a} \ln(p_n) = 0_{\mathbb{H}}$ $(a > 0)$.

Exercise 5.138. *Prove the above properties.*

Exercise 5.139. *Let* $\{p_n\} = \{n - ni + nj + nk\}$. *Compute the following limits:* (a) $\lim_{n\to\infty} e^{p_n}$;   (b) $\lim_{n\to\infty} e^{\frac{1}{p_n}}$;   (c) $\lim_{n\to\infty}(\pm p_n e^{-p_n})$; (d) $\lim_{n\to\infty} \ln(p_n)$;   (e) $\lim_{n\to\infty} \dfrac{\ln(p_n)}{p_n}$;   (f) $\lim_{n\to\infty} p_n^{-3} \ln(p_n)$.

Solution. (a) ∞;   (b) $0_{\mathbb{H}}$;   (c) $0_{\mathbb{H}}$;   (d) ∞;   (e) $0_{\mathbb{H}}$;   (f) $0_{\mathbb{H}}$.

Exercise 5.140. *Let* $\{p_n\} = \{\frac{1}{n} - \frac{1}{n+1}i + \frac{1}{n+2}k\}$. *Compute the following limits:* (a) $\lim_{n\to\infty} \ln(p_n)$;   (b) $\lim_{n\to\infty} e^{p_n}$.

Solution. (a) $-\infty$;   (b) $1_{\mathbb{H}}$.

Exercise 5.141. *Let* $\{p_n\} = \left\{\frac{n}{n+1} - \frac{n^2}{n^2+3}i + \frac{2n^2+1}{n^2+4}k\right\}$. *Compute the following limits:* (a) $\lim_{n\to\infty} \ln(p_n)$;   (b) $\lim_{n\to\infty} e^{p_n}$.

Solution. (a) $\ln(1 - i + 2k)$;   (b) e^{1-i+2k}.

5.5   Principal Value of a Quaternion Natural Logarithm Function

If we wish to work with single-valued branches of $\ln(p)$, it is more satisfactory to restrict $\arg(p)$ to its principal value $\mathrm{Arg}(p)$. This yields the principal branch or principal value of the quaternion logarithm,[3] denoted by the symbol $\mathrm{Ln}(p)$, and defined as

$$\mathrm{Ln}(p) := \log_e |p| + \mathrm{sgn}(\mathbf{p})\,\mathrm{Arg}(p).$$

---

[3] The branch corresponding to $n = 0$ is known as the principal branch, therefore the values that the function takes along this branch are known as principal values.

It is therefore perfectly legitimate to write the principal value of the quaternion logarithm as: $\mathrm{Ln}(p) = \log_e |p| + \mathrm{sgn}(\mathbf{p})\,\theta$, where θ is the quaternion argument of p such that $\theta \in [0, \pi]$.

The principal value of the quaternion logarithm function retains the following properties:

(i) $\mathrm{Ln}(1) = 0$, $\mathrm{Ln}(i) = \frac{\pi}{2}i$, $\mathrm{Ln}(j) = \frac{\pi}{2}j$, and $\mathrm{Ln}(k) = \frac{\pi}{2}k$;

(ii) $\mathrm{Ln}(pq) \neq \mathrm{Ln}(p) + \mathrm{Ln}(q)$ in general, unless p and q commute;

(iii) $\mathrm{Ln}(p^n) = n\mathrm{Ln}(p)$, for $n = 0, \pm 1, \pm 2, \ldots$ (de Moivre's formula);

(iv) For $|p| \geq 1$, $|\mathrm{Ln}(p)| \leq |p| - 1 + \pi$;

(v) For $|p| \geq 1$, $|\mathrm{Ln}(p)| \leq \sum_{k=1}^{2n-1} \frac{(|p|-1)^k}{k} + \pi$, $\quad n \in \mathbb{N}$.

A detailed verification of Properties (i) and (iii) is left to the reader. For (ii), $p = i$, $q = j$; then take for example $\mathrm{Ln}(ij) = \mathrm{Ln}(k) = \frac{\pi}{2}k$, but $\mathrm{Ln}(i) + \mathrm{Ln}(j) = \frac{\pi}{2}(i + j)$. For the proof of (iv), a straightforward computation shows that

$$|\mathrm{Ln}(p)| \leq |\log_e(1 + |p| - 1)| + |\mathrm{sgn}(\mathbf{p})\,\mathrm{Arg}(p)|$$

$$\leq |p| - 1 + \pi.$$

The last step follows from the standard inequality $\ln(1 + x) \leq x$ for every $x \geq 0$. The proof of (v) is a consequence of the classical inequality

$$\ln(1 + x) \leq \sum_{k=1}^{2n-1} (-1)^{k+1} \frac{x^k}{k},$$

which holds for every $x \geq 0$.

Exercise 5.142. *Find all values of the quaternion natural logarithm function at the given point:* (a) $p = \mathrm{Ln}(1)$;   (b) $p = \mathrm{Ln}(2j - 5k)$; $p = \mathrm{Ln}(e^i)$.

Solution. (a) $0_{\mathbb{H}}$;   (b) $\log_e \sqrt{29} + \frac{2j - 5k}{\sqrt{29}} \frac{\pi}{2}$;   (c) i.

Exercise 5.143. *Prove the following relations:* (a) $\mathrm{Ln}(-3) = 2\mathrm{Ln}(i + j + k)$; (b) $3\mathrm{Ln}(i + j + k) = \mathrm{Ln}(-3i - 3j - 3k)$;   (c) $\mathrm{Ln}(9i + 9j + 9k) = 5\mathrm{Ln}(i + j + k)$.

Exercise 5.144. *Let p be an arbitrary real quaternion. Prove the validity of the following statements:* (a) $\mathrm{Ln}(-p) = \mathrm{Ln}(p)$;   (b) $\mathrm{Ln}((-p)^2) = 2\mathrm{Ln}(p)$; (c) $\mathrm{Ln}(\frac{1}{p}) = -\mathrm{Ln}(p)$.

Exercise 5.145. *Show that* $\frac{1}{2} - \frac{\pi}{4} \leq |\mathrm{Ln}(\sqrt{3} + i + j + k)| \leq 1 + \frac{\pi}{4}$.

Exercise 5.146. *Prove that the principal value of the quaternion logarithm satisfies the following estimates:*

$$1 - \frac{1}{|p|} - \arctan\frac{|\mathbf{p}|}{|p_0|} \leq |\mathrm{Ln}(p)| \leq |p| - 1 + \arctan\frac{|\mathbf{p}|}{|p_0|},$$

for any quaternion p.

Exercise 5.147. *Let* $p = i - j + k$. *Prove that:* (a) $|\mathrm{Ln}(p)| \leq 4\sqrt{3} - \frac{14}{3} + \pi$; (b) $|\mathrm{Ln}(p)| \leq \sqrt{3} - 1 + \pi$.

Exercise 5.148. *Let* p *be an arbitrary real quaternion such that* $|p| \geq 2$. *Prove that* $|\mathrm{Ln}(p + i)| \leq \sum_{k=1}^{2n-1} \frac{|p|^k}{k} + \pi$.

5.6   The Inverse of the Quaternion Natural Logarithm Function

At this stage we return to the principle mentioned at the beginning of the chapter. Since $\mathrm{Ln}(p)$ is one of the values of the quaternion logarithm $\ln(p)$, (5.2) indicates that $e^{\mathrm{Ln}(p)} = p$, and $\mathrm{Ln}(e^p) = p$ for all nonzero quaternions p in the so-called fundamental region: $0 < p_0 < +\infty$ and $|\mathbf{p}| < \pi$. This suggests that the quaternion function $\mathrm{Ln}(p)$ plays the role of an inverse function for the exponential quaternion function e^p. To justify this claim, observe that $|e^p| = e^{p_0}$ and $\arg(e^p) = |\mathbf{p}| + 2\pi n$, for $n = 0, \pm 1, \pm 2, \ldots$. Since p is in the fundamental region, it is clear that $\mathrm{Arg}(e^p) = |\mathbf{p}|$. Obviously,

$$e^{\mathrm{Ln}(p)} = e^{\log_e |p| + \mathrm{sgn}(\mathbf{p})\,\mathrm{Arg}(p)}$$

$$= e^{\log_e |p|}\, e^{\mathrm{sgn}(\mathbf{p})\,\mathrm{arccos}\left(\frac{p_0}{|p|}\right)}$$

$$= |p| \left\{ \cos\arccos\left(\frac{p_0}{|p|}\right) + \mathrm{sgn}(\mathbf{p})\sin\arccos\left(\frac{p_0}{|p|}\right) \right\}$$

$$= |p| \left\{ \frac{p_0}{|p|} + \mathrm{sgn}(\mathbf{p})\frac{|\mathbf{p}|}{|p|} \right\} = p.$$

To prove the other relation, note that

$$\mathrm{Ln}(e^p) = \mathrm{Ln}\left\{ e^{p_0}\left(\cos|\mathbf{p}| + \mathrm{sgn}(\mathbf{p})\sin|\mathbf{p}| \right) \right\}$$

$$= \log_e(e^{p_0}) + \mathrm{sgn}(\mathbf{p})\arccos\left(\frac{e^{p_0}\cos|\mathbf{p}|}{e^{p_0}} \right)$$

$$= p_0 + \frac{\mathbf{p}}{|\mathbf{p}|}|\mathbf{p}| = p.$$

The next example shows that there exists a real quaternion p such that $e^{\mathrm{Ln}(p)} = p$, but $\mathrm{Ln}(e^p) = p$ only holds if p is in the fundamental region $0 < p_0 < +\infty$ and

$|\mathbf{p}| < \pi$. For the quaternion $p = \pi k$, which is clearly not in the fundamental region, we have: $e^{\mathrm{Ln}(\pi k)} = e^{\log_e |\pi|} k = \pi k$, but $\mathrm{Ln}(e^{\pi k}) = \mathrm{Ln}(-1) = 0_{\mathbb{H}}$.

Exercise 5.149. *Compute the principal value of the quaternion logarithm* $\mathrm{Ln}(p)$ *for:* (a) $p = (i - (ijk)^3)$; (b) $p = (i + j + k)e^{i+j+k}$; (c) $p = (e^i e^j e^k)^{\mathrm{Sc}(e^{ijk})}$.

Solution. (a) $\log_e \sqrt{2} + i\frac{\pi}{4}$; (b) $\log_e \sqrt{3} + (i + j + k)\arccos(-\sin\sqrt{3})$; (c) $-\frac{\pi}{2}e^{\frac{3}{2}}$.

If $z := x + iy$ is a complex variable then one express z^α with α a complex number, for example, as $e^{\alpha \ln(z)}$. Let us see whether a corresponding statement using the exponential and logarithm quaternion functions holds true.

5.7 Quaternion Power Function

If q is a real quaternion and $p \neq 0_{\mathbb{H}}$, then the quaternion power function p^q is defined to be:

$$p^q := e^{\ln(p)q}. \tag{5.3}$$

This definition certainly holds for $q = 0_{\mathbb{H}}$, so only the case of nonvanishing q has to be considered. The multiple-valuedness of $\arg(p)$ implies that p^q may assume an infinite number of values. Depending on q the quaternion power function will have either one, finitely many, or infinitely many values: if $q = n$ is an integer, then p^n assumes one value only. If $q = \frac{a}{b}$ is a rational number, where a and b are coprime, then $p^{\frac{a}{b}} = |p|^{\frac{a}{b}} e^{\mathrm{sgn}(\mathbf{p})\arg(p)\frac{a}{b}}$ has a finite number of values. If q is a nonzero real quaternion, then p^q always has an infinite number of values.

Quaternion powers satisfy the following properties:

(i) $(p^q)^n = p^{nq}$ for $n = 0, \pm 1, \pm 2, \ldots$;

(ii) $p^{q_1} p^{q_2} \neq p^{q_1+q_2}$ in general, unless $\ln(p)q_1$ and $\ln(p)q_2$ commute.

Example. Let p be a pure quaternion and $q = \frac{2}{3}$. We set

$$p^{\frac{2}{3}} = |p|^{\frac{2}{3}} e^{\mathrm{sgn}(\mathbf{p})\left(\frac{\pi}{2}+2\pi n\right)\frac{2}{3}}$$

$$= |\mathbf{p}|^{\frac{2}{3}} \left[\cos(\frac{\pi}{3} + \frac{4}{3}\pi n) + \mathrm{sgn}(\mathbf{p})\sin(\frac{\pi}{3} + \frac{4}{3}\pi n)\right] \quad (n = 0, 1).$$

Exercise 5.150. *Let p and q be two real quaternions such that $|\mathbf{p}| \neq 0$, and*

$$A := q_0 \log_e |p| - \frac{\mathbf{p} \cdot \mathbf{q}}{|\mathbf{p}|}\left(-\arcsin\frac{p_0}{|p|} + \pi\frac{(4n + 1)}{2}\right),$$

$$B := \frac{\mathbf{p}}{|\mathbf{p}|}\left(-\arcsin\frac{p_0}{|p|} + \pi\frac{4n+1}{2}\right)q_0$$

$$+ \log_e|p|\mathbf{q} + \frac{\mathbf{p}\times\mathbf{q}}{|\mathbf{p}|}\left(-\arcsin\frac{p_0}{|p|} + \pi\frac{4n+1}{2}\right),$$

for $n \in \mathbb{Z}$. Show that

$$p^q = e^A\left(\cos|B| + \text{sgn}(B)\sin|B|\right).$$

Exercise 5.151. *For any nonzero real quaternion p, evaluate $p^{0_{\mathbb{H}}}$.*

Exercise 5.152. *Calculate all possible values of i^i.*

Solution. $e^{-\frac{\pi}{2}}$.

We next turn to some of the more common properties of the above-mentioned elementary quaternion functions, which as in the real and complex cases, should be familiar to the reader.

5.8   General Properties of the Quaternion Power Function

Let $p \in \mathbb{H}$, and $\{p_\nu\}_{\nu=1}^k$ ($k \in \mathbb{N}$) a sequence of elements of $\mathbb{H}$. For $n \in \mathbb{N}$ we have:

(i) $\displaystyle\lim_{n\to\infty} n\left(\sqrt[n]{p} - 1\right) = \log_e|p|$;

(ii) $\displaystyle\lim_{n\to\infty}\left(\frac{1 + \sqrt[n]{p}}{2}\right)^n = \sqrt{|p|}$;

(iii) $\displaystyle\lim_{n\to\infty}\left(\frac{1}{k}\sum_{\nu=1}^k \sqrt[n]{p_\nu}\right)^n = \sqrt[k]{\prod_{\nu=1}^k |p_\nu|}$.

Using the previous definitions, a straightforward computation shows that

$$\sqrt[n]{p} = e^{\frac{1}{n}\log_e|p|}\left[\cos\left|\frac{1}{n}\text{sgn}(\mathbf{p})\arg(p)\right|\right.$$
$$\left. + \text{sgn}\left(\frac{1}{n}\text{sgn}(\mathbf{p})\arg(p)\right)\sin\left|\frac{1}{n}\text{sgn}(\mathbf{p})\arg(p)\right|\right]$$

$$= e^{\frac{1}{n}\log_e|p|}\left[\cos\left|\frac{\arg(p)}{n}\right| + \text{sgn}(\mathbf{p})\sin\left|\frac{\arg(p)}{n}\right|\right].$$

This yields

$$\lim_{n\to\infty} n\left(\sqrt[n]{p} - 1\right) = \lim_{n\to\infty}\frac{e^{\frac{1}{n}\log_e|p|}\left[\cos\left|\frac{\arg(p)}{n}\right| + \text{sgn}(\mathbf{p})\sin\left|\frac{\arg(p)}{n}\right|\right] - 1}{\frac{1}{n}}$$

$$= \log_e|p|.$$

For the proof of (ii), one obviously has

$$\frac{1 + \sqrt[n]{p}}{2} = 1 + \frac{n\left(\sqrt[n]{p} - 1\right)}{2n}.$$

Hence, in accordance with the previous result,

$$\lim_{n\to\infty} \left(\frac{1 + \sqrt[n]{p}}{2}\right)^n = \lim_{n\to\infty} \left(1 + \frac{n(\sqrt[n]{p} - 1)}{2n}\right)^n = e^{\frac{\log_e |p|}{2}} = \sqrt{|p|}.$$

For Property (iii), we clearly have

$$\frac{1}{k} \sum_{v=1}^{k} \sqrt[n]{p_v} = 1 + \frac{1}{kn}\left[n \sum_{v=1}^{k}\left(\sqrt[n]{p_v} - 1\right)\right],$$

and, as the above discussion shows,

$$\lim_{n\to\infty} \frac{1}{k}\left[n \sum_{v=1}^{k}\left(\sqrt[n]{p_v} - 1\right)\right] = \frac{1}{k} \sum_{v=1}^{k} \log_e |p_v| = \log_e \sqrt[k]{\prod_{v=1}^{k} |p_v|}.$$

With these calculations at hand, we set

$$\lim_{n\to\infty} \left(\frac{1}{k} \sum_{v=1}^{k} \sqrt[n]{p_v}\right)^n = e^{\log_e\left(\sqrt[k]{\prod_{v=1}^{k} |p_v|}\right)} = \sqrt[k]{\prod_{v=1}^{k} |p_v|}.$$

Example. Prove that $\lim_{n\to\infty} \frac{1}{2^n}\left(\sqrt[n^2]{p} + \sqrt[n^3]{p}\right)^n = 1_{\mathbb{H}}$.

Solution. Direct computations show that

$$\sqrt[n^2]{p} = e^{\frac{1}{n^2} \log_e |p|}\left[\cos\left|\frac{\arg(p)}{n^2}\right| + \operatorname{sgn}(\mathbf{p}) \sin\left|\frac{\arg(p)}{n^2}\right|\right],$$

$$\sqrt[n^3]{q} = e^{\frac{1}{n^3} \log_e |q|}\left[\cos\left|\frac{\arg(q)}{n^3}\right| + \operatorname{sgn}(\mathbf{p}) \sin\left|\frac{\arg(q)}{n^3}\right|\right].$$

It follows that

$$\lim_{n\to\infty} n\left(\sqrt[n^2]{p} - 1\right) = 0_{\mathbb{H}} \quad \text{and} \quad \lim_{n\to\infty} n\left(\sqrt[n^3]{q} - 1\right) = 0_{\mathbb{H}}.$$

Consequently,

$$\lim_{n\to\infty} \frac{\left(\sqrt[n^2]{p} + \sqrt[n^3]{q}\right)^n}{2^n} = \lim_{n\to\infty} \left[1 + \frac{n\left(\sqrt[n^2]{p} - 1\right) + n\left(\sqrt[n^3]{q} - 1\right)}{2n}\right]^n$$

$$= 1_{\mathbb{H}}.$$

Exercise 5.153. *Prove that:*
(a) $\displaystyle\lim_{n\to\infty} n((\sqrt{3} = i + j + k)^{\frac{1}{n}} - 1) = \log_e 2;$

(b) $\displaystyle\lim_{n\to\infty} \left(\frac{1 + (\sqrt{3} + i + j + k)^{\frac{1}{n}}}{2}\right)^n = \sqrt{2};$

(c) $\displaystyle\lim_{n\to\infty} \frac{\left((\sqrt{3} + i + j + k)^{\frac{1}{n^2}} + (\sqrt{3} + i + j + k)^{\frac{1}{n^3}}\right)^n}{2^n} = 1_{\mathbb{H}}.$

Exercise 5.154. *Let $\{p_n\}$ and $\{q_n\}$ ($n \in \mathbb{N}$) be two sequences of elements of $\mathbb{H}$ such that $\lim_{n\to\infty} p_n = p$ and $\lim_{n\to\infty} q_n = q$. Prove that*

$$\lim_{n\to\infty} \frac{1}{2^n}\left(\sqrt[n]{p_n} + \sqrt[n]{q_n}\right)^n = \sqrt{|p||q|}.$$

5.9　Principal Value of a Quaternion Power Function

We can assign a unique value to p^q by using the principal value of the quaternion logarithm $\mathrm{Ln}(p)$ instead of $\ln(p)$. This particular value is called the principal value of the quaternion power function p^q, and is defined by

$$p^q := e^{\mathrm{Ln}(p)q}.$$

Exercise 5.155. *Find the principal value of the given quaternion powers:* (a) i^i; (b) i^j;　(c) i^k;　(d) $(i + j + k)^{i+j+k}$.

Solution. (a) $e^{-\frac{\pi}{2}}$;　(b) k;　(c) $-j$;　(d) $e^{-\pi\frac{\sqrt{3}}{2}}$.

Exercise 5.156. *Let α be a real number. Under what circumstances does the property $p^\alpha q^\alpha = (pq)^\alpha$ hold for quaternion numbers? And, does the property $p^\alpha q^\alpha = (pq)^\alpha$ hold for the principal value of a quaternion power?*

5.10　Advanced Practical Exercises

1. Compute e^p where (a) $p = 1 - i + j$;　(b) $p = 2 + i - j$;　(c) $p = i - k$;　(d) $p = i$;　(e) $p = j$;　(f) $p = k$.

2. Prove the validity of the following relations:

$$(1)\ e^{-i}\frac{1+k}{2} = \frac{1+k}{2}e^{j};\quad (2)\ e^{i}\frac{1+k}{2} = \frac{1+k}{2}e^{-j};\quad (3)\ e^{-i}\frac{1-k}{2} = \frac{1-k}{2}e^{-j};$$

$$(4)\ e^{i}\frac{1-k}{2} = \frac{1-k}{2}e^{j};\quad (5)\ e^{i}\frac{1+j}{2} = \frac{1+j}{2}e^{k};\quad (6)\ e^{i}\frac{1-j}{2} = \frac{1-j}{2}e^{-k};$$

$$(7)\ e^{i}\frac{1+j}{2} = \frac{1+j}{2}e^{k};\quad (8)\ e^{i}\frac{1-k}{2} = \frac{1-k}{2}e^{j};\quad (9)\ e^{i}\frac{1+i}{2} = \frac{1+i}{2}e^{i};$$

$$(10)\ e^{i}\frac{1-i}{2} = \frac{1-i}{2}e^{i};\quad (11)\ e^{-i}\frac{1+i}{2} = \frac{1+i}{2}e^{-i};\quad (12)\ e^{-i}\frac{1-i}{2} = \frac{1-i}{2}e^{-i};$$

$$(13)\ e^{j}\frac{1+i}{2} = \frac{1+i}{2}e^{-k};\quad (14)\ e^{j}\frac{1-i}{2} = \frac{1-i}{2}e^{k};\quad (15)\ e^{j}\frac{1+j}{2} = \frac{1+j}{2}e^{j};$$

$$(16)\ e^{j}\frac{1-j}{2} = \frac{1-j}{2}e^{j};\quad (17)\ e^{j}\frac{1+k}{2} = \frac{1+k}{2}e^{i};\quad (18)\ e^{j}\frac{1-k}{2} = \frac{1-k}{2}e^{-i};$$

$$(19)\ e^{k}\frac{1-k}{2} = \frac{1-k}{2}e^{k};\quad (20)\ e^{k}\frac{1+k}{2} = \frac{1+k}{2}e^{k};\quad (21)\ e^{k}\frac{1+i}{2} = \frac{1+i}{2}e^{j};$$

$$(22)\ e^{k}\frac{1-i}{2} = \frac{1-i}{2}e^{-j};\quad (23)\ e^{k}\frac{1+j}{2} = \frac{1+j}{2}e^{-i};\quad (24)\ e^{k}\frac{1-j}{2} = \frac{1-j}{2}e^{i};$$

$$(25)\ e^{-i}\frac{1+j}{2} = \frac{1+j}{2}e^{-k};\quad (26)\ e^{-i}\frac{1+k}{2} = \frac{1+k}{2}e^{j};\quad (27)\ e^{-i}\frac{1-j}{2} = \frac{1-j}{2}e^{k};$$

$$(28)\ e^{-i}\frac{1-k}{2} = \frac{1-k}{2}e^{-j};\quad (29)\ e^{-j}\frac{1-j}{2} = \frac{1-j}{2}e^{-j};\quad (30)\ e^{-j}\frac{1+j}{2} = \frac{1+j}{2}e^{-j};$$

$$(31)\ e^{-j}\frac{1+i}{2} = \frac{1+i}{2}e^{k};\quad (32)\ e^{-j}\frac{1+k}{2} = \frac{1+k}{2}e^{-i};\quad (33)\ e^{-j}\frac{1-i}{2} = \frac{1-i}{2}e^{-k};$$

$$(34)\ e^{-j}\frac{1-k}{2} = \frac{1-k}{2}e^{-j};\quad (35)\ e^{-k}\frac{1+i}{2} = \frac{1+i}{2}e^{-j};\quad (36)\ e^{-k}\frac{1-i}{2} = \frac{1-i}{2}e^{j};$$

$$(37)\ e^{-k}\frac{1+j}{2} = \frac{1+j}{2}e^{i};\quad (38)\ e^{-k}\frac{1-j}{2} = \frac{1-j}{2}e^{-i};\quad (39)\ e^{-k}\frac{1-k}{2} = \frac{1-k}{2}e^{-k};$$

$$(40)\ e^{-k}\frac{1+k}{2} = \frac{1+k}{2}e^{-k}.$$

3. Let $p = a_1 + b_1 i + c_1 j + d_1 k$ and $q = a_2 + b_2 i + c_2 j + d_2 k$, where $a_l, b_l, c_l, d_l \in \mathbb{R}$ $(l = 1, 2)$. Prove the validity of the following relations:

$$(1)\ e^{-i(p \cdot q)}\frac{1+k}{2} = \frac{1+k}{2}e^{j(p \cdot q)};\quad (2)\ e^{i(p \cdot q)}\frac{1+k}{2} = \frac{1+k}{2}e^{-j(p \cdot q)};$$

$$(3)\ e^{-i(p \cdot q)}\frac{1-k}{2} = \frac{1-k}{2}e^{-j(p \cdot q)};\quad (4)\ e^{i(p \cdot q)}\frac{1-k}{2} = \frac{1-k}{2}e^{j(p \cdot q)};$$

$$(5)\ e^{i(p \cdot q)}\frac{1+j}{2} = \frac{1+j}{2}e^{k(p \cdot q)};\quad (6)\ e^{i(p \cdot q)}\frac{1-j}{2} = \frac{1-j}{2}e^{-k(p \cdot q)};$$

$$(7)\ e^{i(p \cdot q)}\frac{1+j}{2} = \frac{1+j}{2}e^{k(p \cdot q)};\quad (8)\ e^{i(p \cdot q)}\frac{1-k}{2} = \frac{1-k}{2}e^{j(p \cdot q)};$$

$(9)\ e^{i(p\cdot q)}\dfrac{1+i}{2} = \dfrac{1+i}{2}e^{i(p\cdot q)}; \qquad (10)\ e^{i(p\cdot q)}\dfrac{1-i}{2} = \dfrac{1-i}{2}e^{i(p\cdot q)};$

$(11)\ e^{-i(p\cdot q)}\dfrac{1+i}{2} = \dfrac{1+i}{2}e^{-i(p\cdot q)}; \qquad (12)\ e^{-i(p\cdot q)}\dfrac{1-i}{2} = \dfrac{1-i}{2}e^{-i(p\cdot q)};$

$(13)\ e^{j(p\cdot q)}\dfrac{1+i}{2} = \dfrac{1+i}{2}e^{-k(p\cdot q)}; \qquad (14)\ e^{j(p\cdot q)}\dfrac{1-i}{2} = \dfrac{1-i}{2}e^{k(p\cdot q)};$

$(15)\ e^{j(p\cdot q)}\dfrac{1+j}{2} = \dfrac{1+j}{2}e^{j(p\cdot q)}; \qquad (16)\ e^{j(p\cdot q)}\dfrac{1-j}{2} = \dfrac{1-j}{2}e^{j(p\cdot q)};$

$(17)\ e^{j(p\cdot q)}\dfrac{1+k}{2} = \dfrac{1+k}{2}e^{i(p\cdot q)}; \qquad (18)\ e^{j(p\cdot q)}\dfrac{1-k}{2} = \dfrac{1-k}{2}e^{-i(p\cdot q)};$

$(19)\ e^{k(p\cdot q)}\dfrac{1-k}{2} = \dfrac{1-k}{2}e^{k(p\cdot q)}; \qquad (20)\ e^{k(p\cdot q)}\dfrac{1+k}{2} = \dfrac{1+k}{2}e^{k(p\cdot q)};$

$(21)\ e^{k(p\cdot q)}\dfrac{1+i}{2} = \dfrac{1+i}{2}e^{j(p\cdot q)}; \qquad (22)\ e^{k(p\cdot q)}\dfrac{1-i}{2} = \dfrac{1-i}{2}e^{-j(p\cdot q)};$

$(23)\ e^{k(p\cdot q)}\dfrac{1+j}{2} = \dfrac{1+j}{2}e^{-i(p\cdot q)}; \qquad (24)\ e^{k(p\cdot q)}\dfrac{1-j}{2} = \dfrac{1-j}{2}e^{i(p\cdot q)};$

$(25)\ e^{-i(p\cdot q)}\dfrac{1+j}{2} = \dfrac{1+j}{2}e^{-k(p\cdot q)}; \qquad (26)\ e^{-i(p\cdot q)}\dfrac{1+k}{2} = \dfrac{1+k}{2}e^{j(p\cdot q)};$

$(27)\ e^{-i(p\cdot q)}\dfrac{1-j}{2} = \dfrac{1-j}{2}e^{k(p\cdot q)}; \qquad (28)\ e^{-i(p\cdot q)}\dfrac{1-k}{2} = \dfrac{1-k}{2}e^{-j(p\cdot q)};$

$(29)\ e^{-j(p\cdot q)}\dfrac{1-j}{2} = \dfrac{1-j}{2}e^{-j(p\cdot q)}; \qquad (30)\ e^{-j(p\cdot q)}\dfrac{1+j}{2} = \dfrac{1+j}{2}e^{-j(p\cdot q)};$

$(31)\ e^{-j(p\cdot q)}\dfrac{1+i}{2} = \dfrac{1+i}{2}e^{k(p\cdot q)}; \qquad (32)\ e^{-j(p\cdot q)}\dfrac{1+k}{2} = \dfrac{1+k}{2}e^{-i(p\cdot q)};$

$(33)\ e^{-j(p\cdot q)}\dfrac{1-i}{2} = \dfrac{1-i}{2}e^{-k(p\cdot q)}; \qquad (34)\ e^{-j(p\cdot q)}\dfrac{1-k}{2} = \dfrac{1-k}{2}e^{-j(p\cdot q)};$

$(35)\ e^{-k(p\cdot q)}\dfrac{1+i}{2} = \dfrac{1+i}{2}e^{-j(p\cdot q)}; \qquad (36)\ e^{-k(p\cdot q)}\dfrac{1-i}{2} = \dfrac{1-i}{2}e^{j(p\cdot q)};$

$(37)\ e^{-k(p\cdot q)}\dfrac{1+j}{2} = \dfrac{1+j}{2}e^{i(p\cdot q)}; \qquad (38)\ e^{-k(p\cdot q)}\dfrac{1-j}{2} = \dfrac{1-j}{2}e^{-i(p\cdot q)};$

$(39)\ e^{-k(p\cdot q)}\dfrac{1-k}{2} = \dfrac{1-k}{2}e^{-k(p\cdot q)}; \qquad (40)\ e^{-k(p\cdot q)}\dfrac{1+k}{2} = \dfrac{1+k}{2}e^{-k(p\cdot q)}.$

4. Let $a,b \in \mathbb{R}$. Prove that $e^{-i(a\mp b)}\frac{1\pm k}{2} = \frac{1\pm k}{2}e^{-i(b\mp a)}$.

5. Let $a,b \in \mathbb{R}$ and $p = e^{-ia}e^{-jb}$. Prove that $p_+ = e^{-i(a-b)}\frac{1+k}{2}$ and $p_- = e^{-i(a+b)}\frac{1-k}{2}$.

6. Compute (a) $pq - \bar{r} + e^{r} + [p,q];$ (b) $q - \overline{pr} + p\cdot q - (p,q) + e^{r};$ (c) $p - q + \arg(r)p + \overline{e^{p}}$, where $p = 1 + i + j,\, q = i + k$, and $r = j + k$.

7. Compute $\ln(p)$ where (a) $p = 1 + i + j + k$; (b) $p = i$; (c) $p = j$; (d) $p - k$; (e) $p = i + j$; (f) $p = i + k$; (g) $p = j + k$.

8. Find the first and second matrix representation of $\ln(p)$ if (a) $p = 1 - i - j + k$; (b) $p = 2 - 3i + \sqrt{3}j$; (c) $p = 1 + k$.

9. Compute (a) $\ln(p) + \arg(q)\ln(r) + [p, q] - p \cdot r$; (b) $\ln(q) - (p, r)$; (c) $\ln(p) - \ln(q) + [p, q + r]$, where $p = 1 - i + j - k$, $q = i + j$, and $r = i + k$.

10. Let p be a real quaternion such that $|\mathbf{p}| \neq 0$ and $p_0 \geq 0$. Prove that

$$\ln(p) = \log_e |p| + \text{sgn}(\mathbf{p})\left(\arcsin \frac{|\mathbf{p}|}{|p|} + 2n\pi\right), \quad n \in \mathbb{Z}.$$

11. Let p be a real quaternion such that $|\mathbf{p}| \neq 0$ and $p_0 \leq 0$. Prove that

$$\ln(p) = \log_e |p| + \text{sgn}(\mathbf{p})\left((2n + 1)\pi - \arcsin \frac{|\mathbf{p}|}{|p|}\right), \quad n \in \mathbb{Z}.$$

12. Let p be a real quaternion such that $|\mathbf{p}| \neq 0$ and $p_0 < 0$. Prove that

$$\ln(p) = \log_e |p| + \text{sgn}(\mathbf{p})\left((2n + 1)\pi + \arctan \frac{|\mathbf{p}|}{p_0}\right), \quad n \in \mathbb{Z}.$$

13. Let p be a real quaternion such that $|\mathbf{p}| \neq 0$ and $p_0 > 0$. Prove that

$$\ln(p) = \log_e |p| + \text{sgn}(\mathbf{p})\left(\arctan \frac{p_0}{|\mathbf{p}|} + 2n\pi\right), \quad n \in \mathbb{Z}.$$

14. Compute (a) $(1 + i)^2$; (b) $(1 - i - j + k)^4$; (c) $(1 + i + j + k)^{10}$.

15. Compute (a) $2^{1-i+j+k}$; (b) 3^{1-i}; (c) 4^{i+j+k}; (d) e^{1+j}; (e) e^{1+i+j}; (f) $e^{1+i+j+k}$.

16. Compute (a) i^j; (b) i^k; (c) j^k; (d) $(i+j)^k$; (e) $(i+k)^j$; (f) $(j+k)^i$.

17. Find the first matrix representation of the following quaternion

$$j^k(1 + i - j) + \overline{2 - i - 3j + k} + [i + j, i - j].$$

18. Compute $j^k \frac{1-i-k}{j+k}$.

19. Let $p = a_1 + b_1 i + c_1 j + d_1 k$, $q = a_2 + b_2 i + c_2 j + d_2 k$, where $a_l, b_l, c_l, d_l \in \mathbb{R}$ ($l = 1, 2$). Find p^q.

20. Compute $A = i^j + j^k + i^k + k^j + k^i + j^i$.

21. Let p and q be two real quaternions such that $|\mathbf{p}| \neq 0$, $p_0 \geq 0$, and

$$A := q_0 \log_e |p| - \frac{\mathbf{p} \cdot \mathbf{q}}{|\mathbf{p}|}\left(\arcsin \frac{|\mathbf{p}|}{|p|} + 2n\pi\right),$$

$$B := \frac{\mathbf{p}}{|\mathbf{p}|}\left(\arcsin \frac{|\mathbf{p}|}{|p|} + 2n\pi\right)q_0 + \log_e |p|\mathbf{q} + \frac{\mathbf{p} \times \mathbf{q}}{|\mathbf{p}|}\left(\arcsin \frac{|\mathbf{p}|}{|p|} + 2n\pi\right)$$

for $n \in \mathbb{Z}$. Prove that

$$p^q \;=\; e^A\Big(\cos|B| + \mathrm{sgn}(B)\sin|B|\Big).$$

22. Let p and q be two real quaternions such that $|\mathbf{p}| \neq 0$, $p_0 \leq 0$, and

$$A := q_0 \log_e |p| - \frac{\mathbf{p}\cdot\mathbf{q}}{|\mathbf{p}|}\Big(-\arcsin\frac{|\mathbf{p}|}{|p|} + (2n+1)\pi\Big),$$

$$B := \frac{\mathbf{p}}{|\mathbf{p}|}\Big(-\arcsin\frac{|\mathbf{p}|}{|p|} + (2n+1)\pi\Big)q_0$$

$$+ \log_e|p|\mathbf{q} + \frac{\mathbf{p}\times\mathbf{q}}{|\mathbf{p}|}\Big(-\arcsin\frac{|\mathbf{p}|}{|p|} + (2n+1)\pi\Big),$$

for $n \in \mathbb{Z}$. Prove that

$$p^q \;=\; e^A\Big(\cos|B| + \mathrm{sgn}(B)\sin|B|\Big).$$

23. Let p and q be two real quaternions such that $|\mathbf{p}| \neq 0$, $p_0 < 0$, and

$$A := q_0 \log_e |p| - \frac{\mathbf{p}\cdot\mathbf{q}}{|\mathbf{p}|}\Big(\arctan\frac{|\mathbf{p}|}{p_0} + (2n+1)\pi\Big),$$

$$B := \frac{\mathbf{p}}{|\mathbf{p}|}\Big(\arctan\frac{|\mathbf{p}|}{p_0} + (2n+1)\pi\Big)q_0$$

$$+ \log_e|p|\mathbf{q} + \frac{\mathbf{p}\times\mathbf{q}}{|\mathbf{p}|}\Big(\arctan\frac{|\mathbf{p}|}{p_0} + (2n+1)\pi\Big),$$

for $n \in \mathbb{Z}$. Prove that

$$p^q \;=\; e^A\Big(\cos|B| + \mathrm{sgn}(B)\sin|B|\Big).$$

24. Let p and q be two real quaternions such that $|\mathbf{p}| \neq 0$, $p_0 > 0$, and

$$A := q_0 \log_e |p| - \frac{\mathbf{p}\cdot\mathbf{q}}{|\mathbf{p}|}\Big(\mathrm{arccotan}\frac{p_0}{|\mathbf{p}|} + 2n\pi\Big),$$

$$B := \frac{\mathbf{p}}{|\mathbf{p}|}\Big(\mathrm{arccotan}\frac{p_0}{|\mathbf{p}|} + 2n\pi\Big)q_0$$

$$+ \log_e|p|\mathbf{q} + \frac{\mathbf{p}\times\mathbf{q}}{|\mathbf{p}|}\Big(\mathrm{arccotan}\frac{p_0}{|\mathbf{p}|} + 2n\pi\Big),$$

for $n \in \mathbb{Z}$. Prove that

$$p^q = e^A \Big(\cos |B| + \mathrm{sgn}(B) \sin |B| \Big).$$

25. Let p and q be two real quaternions such that $|\mathbf{p}| \neq 0$, $p_0 > 0$, and

$$A := q_0 \log_e |p| - \frac{\mathbf{p} \cdot \mathbf{q}}{|\mathbf{p}|} \Big(\arctan \frac{|\mathbf{p}|}{p_0} + 2n\pi \Big),$$

$$B := \frac{\mathbf{p}}{|\mathbf{p}|} \Big(\arctan \frac{|\mathbf{p}|}{p_0} + 2n\pi \Big) q_0$$

$$+ \log_e |p| \mathbf{q} + \frac{\mathbf{p} \times \mathbf{q}}{|\mathbf{p}|} \Big(\arctan \frac{|\mathbf{p}|}{p_0} + 2n\pi \Big),$$

for $n \in \mathbb{Z}$. Prove that

$$p^q = e^A \Big(\cos |B| + \mathrm{sgn}(B) \sin |B| \Big).$$

26. Let p and q be two real quaternions such that $|\mathbf{p}| \neq 0$, and

$$A := q_0 \log_e |p| - \frac{\mathbf{p} \cdot \mathbf{q}}{|\mathbf{p}|} \Big(\arccos \frac{p_0}{|p|} + 2n\pi \Big),$$

$$B := \frac{\mathbf{p}}{|\mathbf{p}|} \Big(\arccos \frac{p_0}{|p|} + 2n\pi \Big) q_0$$

$$+ \log_e |p| \mathbf{q} + \frac{\mathbf{p} \times \mathbf{q}}{|\mathbf{p}|} \Big(\arccos \frac{p_0}{|p|} + 2n\pi \Big),$$

for $n \in \mathbb{Z}$. Prove that

$$p^q = e^A \Big(\cos |B| + \mathrm{sgn}(B) \sin |B| \Big).$$

Trigonometric Functions 6

In this chapter we define quaternion trigonometric functions. Analogously to the quaternion functions e^p and $\ln(p)$, these functions will agree with their counterparts for real and complex input. In addition, we will show that the quaternion trigonometric functions satisfy many of the same identities the real and complex trigonometric functions do. We further note that some identities will be used in the calculus of these quaternion functions without previous introduction. Even though our aim is to smooth the path for the reader who wishes to gain an understanding of these results in quaternion trigonometry, we still include complete proofs of such identities in this introductory exposition.

With the help of the exponential function (5.1), quaternionic analogues of the trigonometric functions can be introduced.

6.1 Quaternion Sine and Cosine Functions

The functions $\sin(p)$ and $\cos(p)$ defined respectively by

$$\sin(p) := \begin{cases} -\frac{1}{2}\,\mathrm{sgn}(\mathbf{p})\left(e^{p\,\mathrm{sgn}(\mathbf{p})} - e^{-p\,\mathrm{sgn}(\mathbf{p})}\right), & \text{if } |\mathbf{p}| \neq 0 \\ \sin(p_0), & \text{if } |\mathbf{p}| = 0 \end{cases}$$

$$\cos(p) := \begin{cases} \frac{1}{2}\left(e^{p\,\mathrm{sgn}(\mathbf{p})} + e^{-p\,\mathrm{sgn}(\mathbf{p})}\right), & \text{if } |\mathbf{p}| \neq 0 \\ \cos(p_0), & \text{if } |\mathbf{p}| = 0 \end{cases}$$

are called the quaternion sine and cosine functions. It is clear that the quaternion trigonometric functions are natural generalizations of the conventional ones we are familiar with in the real and complex settings. Of course another way of defining the sine and cosine quaternion functions would be via their power series. This leads to the alternative definitions,

J.P. Morais et al., *Real Quaternionic Calculus Handbook*,
DOI 10.1007/978-3-0348-0622-0_6, © Springer Basel 2014

$$\sin(p) = \sum_{l=0}^{\infty} \frac{(-1)^l}{(2l+1)!} \, p^{2l+1} \quad \text{and} \quad \cos(p) = \sum_{l=0}^{\infty} \frac{(-1)^l}{(2l)!} \, p^{2l}.$$

If we write down the series for $e^{\pm p \, \mathrm{sgn}(\mathbf{p})}$ one can easily verify that both approaches yield the same quaternion function. The proof is left to the reader as an exercise.

Exercise 6.157. *Let p be a real quaternion. Prove the validity of the trigonometric identities:* (a) $\overline{\cos(p)} = \cos(\overline{p})$; (b) $\overline{\sin(p)} = -\sin(\overline{p})$.

Exercise 6.158. *Compute the following quaternion power series:*
(a) $\sum_{l=0}^{\infty} \frac{(-1)^l}{(2l+1)!} (i+j+k)^{2l+1}$; (b) $\sum_{l=0}^{\infty} \frac{(-1)^l}{(2l)!} (1-i-j-k)^{2l}$.

Solution. (a) $\sin(i+j+k)$; (b) $\cos(1-i-j-k)$.

Example. Let $p = 1 + i + j + k$. Compute $\cos(p)$.

Solution. First note that $|\mathbf{p}| = \sqrt{3}$ and $\mathrm{sgn}(\mathbf{p}) = \frac{1}{\sqrt{3}}(i+j+k)$. A straightforward computation shows that

$$e^{\pm p \, \mathrm{sgn}(\mathbf{p})} = e^{\mp\sqrt{3}}\left(\cos(1) + \frac{i+j+k}{\sqrt{3}} \sin(1)\right).$$

It follows that[1]

$$\cos(1 + i + j + k) = \cos(1)\cosh(\sqrt{3}) - \frac{i+j+k}{\sqrt{3}} \sin(1)\sinh(\sqrt{3}).$$

Exercise 6.159. *Compute the value of the given quaternion trigonometric function:*
(a) $\sin(1 + i + j + k)$; (b) $\sin(-2i + j)$; (c) $\cos(i^j - j^i)$; (d) $\sin(k) - 7\sin(i+j)\cos(j+k)$; (e) $\sin(\cos(-i+3k))$.

Solution. (a) $\sin(1+i+j+k) = -\sin 1 \sinh \sqrt{3} + \frac{\cos 1 \sinh \sqrt{3}}{\sqrt{3}}(i+j+k)$; (b) $\frac{\sqrt{5}}{5}\sinh\sqrt{5}(-2i+j)$; (c) $k \sinh 1$; (d) $-\frac{7\sqrt{2}}{4}\sinh(2\sqrt{2})(i+j) + k \sinh 1$; (e) $\sin(\sinh\sqrt{10})$.

---

[1] If t is a real variable, then the real hyperbolic sine and cosine functions are defined using the real exponential function as follows:

$$\sinh(t) = \frac{e^t - e^{-t}}{2} \quad \text{and} \quad \cosh(t) = \frac{e^t + e^{-t}}{2}.$$

6.2   Trigonometric Identities

The usual identities, such as $\sin(p) = -\sin(-p)$ and $\cos(p) = \cos(-p)$, hold here as well. This follows from the previous definitions and properties of the quaternion exponential function. We may then ask whether or not other familiar identities for real and complex trigonometric functions hold in a quaternionic context. A straightforward computation shows that for $\sin^2(p) + \cos^2(p) = 1_\mathbb{H}$ the answer is yes. To get a handle on the proof, we first derive the relations

$$p \operatorname{sgn}(\mathbf{p}) = p_0 \operatorname{sgn}(\mathbf{p}) + \mathbf{p} \operatorname{sgn}(\mathbf{p})$$

$$= -|\mathbf{p}| + p_0 \operatorname{sgn}(\mathbf{p})$$

and

$$\operatorname{sgn}\left(p_0 \operatorname{sgn}(\mathbf{p})\right) = \frac{p_0 \operatorname{sgn}(\mathbf{p})}{|p_0|} = \operatorname{sgn}(p_0) \operatorname{sgn}(\mathbf{p}).$$

We can go one step further and deduce the identities

$$e^{p \operatorname{sgn}(\mathbf{p})} = e^{-|\mathbf{p}|} e^{p_0 \operatorname{sgn}(\mathbf{p})}$$

$$= e^{-|\mathbf{p}|}\left(\cos(p_0) + \operatorname{sgn}(\mathbf{p}) \sin(p_0)\right),$$

from which it follows that

$$\sin(p) = \sin(p_0) \cos(\mathbf{p}) + \cos(p_0) \sin(\mathbf{p}),$$

$$\cos(p) = \cos(p_0) \cos(\mathbf{p}) - \sin(p_0) \sin(\mathbf{p}).$$

The rest of the proof can be done without too much difficulty by computation and is left as an exercise for the reader. Let us now give up an example. Take $p = i + j$; then $\sin^2(i + j) + \cos^2(i + j) = \frac{(i+j)^2}{2} \sinh^2(\sqrt{2}) + \cosh^2(\sqrt{2}) = 1_\mathbb{H}$. However, one has to pay attention to the fact that in general the quaternion trigonometric functions do not satisfy the sum and difference formulae $\sin(p \pm q) = \sin p \cos q \pm \cos p \sin q$ and $\cos(p \pm q) = \cos p \cos q \mp \sin p \sin q$, since $\operatorname{sgn}(\mathbf{p} + \mathbf{q})$ is not linear in $\mathbf{p}$ and $\mathbf{q}$, respectively; a fact that often leads to complications. Take for example $p = i + j$; then $\sin(i + j) = \frac{(i+j)}{\sqrt{2}} \sinh(\sqrt{2}) \neq (i + j) \sinh(1) \cosh(1) = \sin(i) \cos(j) + \cos(i) \sin(j)$. Similarly to the complex case, it is also important to realize that some other properties of the real trigonometric functions are not satisfied by their quaternionic counterparts. For example, while $|\sin(t)| \leq 1$ and $|\cos(t)| \leq 1$ for all real t, note that $|\sin(i + j)| \approx 1.935 > 1$ and $|\cos(i + j)| \approx 2.178 > 1$.

Exercise 6.160. *Let $p \in \mathbb{H} \setminus \{0_\mathbb{H}\}$, $p_0 \operatorname{sgn}(\mathbf{p}) = -i - j - k$, and $p \operatorname{sgn}(\mathbf{p}) = -1 - i - j - k$. Find $|p|$.*

Solution. $|p| = 1$.

Exercise 6.161. *Let* $p \in \mathbb{H} \setminus \{0_{\mathbb{H}}\}$, *and* $\mathrm{sgn}(p_0)\mathrm{sgn}(\mathbf{p}) = i + j + k$. *Find* $\mathrm{sgn}(p_0\mathrm{sgn}(\mathbf{p}))$.

Solution. $i + j + k$.

Exercise 6.162. *Find an* $n \in \mathbb{N}$ *such that* $|\sin(nk)| > \frac{3}{2}$.

Solution. $n \geq \log_e \frac{3+\sqrt{10}}{2}$.

Exercise 6.163. *Let* p *and* q *be two real quaternions. Find a counterexample to show that the sum and difference formulae are not necessarily satisfied by quaternion trigonometric functions.*

Solution. Take $p = i + j$ and $q = 1 + k$.

Exercise 6.164. *Let* $p = a + bi + cj + dk$. *Find all quaternions* p *such that* $\sin(p) = \cos(p)$.

Solution. p is such that $a = \frac{\pi}{4} + n\pi$ $(n \in \mathbb{Z})$, and $b = c = d = 0$.

6.3   Trigonometric Equations

Keeping the previous notions, we now turn our attention to quaternion trigonometric equations. It is clear that always there are either no solutions, finitely many, or infinitely many solutions to equations of the form $\sin(p) = w$ or $\cos(p) = w$, for a given quaternion w; this has to do with the possible many-valuedness of these functions. There are several methods to reduce a given equation to an equivalent, but more manageable expression, however we shall focus almost exclusively on using previous definitions together with algebraic techniques. For example, let $p = bi$ be a solution to the equation $\sin(p) = i$. The discussion of this example presents no particular difficulty. A straightforward computation shows that $b \sinh(|b|) = |b|$, and when solved this equation gives $p = \pm \ln(1 + \sqrt{2})i$. Each of these values of p satisfies the equation $\sin(p) = i$.

Example. Let $p = a + bi + cj + dk$. Solve the equation $\cos(p) = 1_{\mathbb{H}}$.

Solution. We first compute $\cos(p)$ explicitly. We have

$$\cos(p) = \cos\left(\sqrt{a^2 + b^2 + c^2 + d^2}\right) \cosh\left(\sqrt{b^2 + c^2 + d^2}\right)$$

$$- (bi + cj + dk) \sin\left(\sqrt{a^2 + b^2 + c^2 + d^2}\right) \sinh\left(\sqrt{b^2 + c^2 + d^2}\right).$$

We obtain the following system for the parameters a, b, c, d:

$$\begin{cases} \cos\left(\sqrt{a^2 + b^2 + c^2 + d^2}\right) \cosh\left(\sqrt{b^2 + c^2 + d^2}\right) = 1, \\[2ex] b \sin\left(\sqrt{a^2 + b^2 + c^2 + d^2}\right) \sinh\left(\sqrt{b^2 + c^2 + d^2}\right) = 0, \\[2ex] c \sin\left(\sqrt{a^2 + b^2 + c^2 + d^2}\right) \sinh\left(\sqrt{b^2 + c^2 + d^2}\right) = 0, \\[2ex] d \sin\left(\sqrt{a^2 + b^2 + c^2 + d^2}\right) \sinh\left(\sqrt{b^2 + c^2 + d^2}\right) = 0. \end{cases}$$

This implies that $b = c = d = 0$, and a is such that $\cos(a) = 1$, and so $a = 2n\pi$ $(n \in \mathbb{Z})$. It follows that $p = 2n\pi$ with $n \in \mathbb{Z}$.

Exercise 6.165. *Let $p = a + bi + cj + dk$. Compute $\sin(p)$.*

Solution. The value is computed as

$$\sin a \cosh \sqrt{b^2 + c^2 + d^2} + \frac{b \cos a \sinh \sqrt{b^2 + c^2 + d^2}}{\sqrt{b^2 + c^2 + d^2}} i$$

$$+ \frac{c \cos a \sinh \sqrt{b^2 + c^2 + d^2}}{\sqrt{b^2 + c^2 + d^2}} j + \frac{d \cos a \sinh \sqrt{b^2 + c^2 + d^2}}{\sqrt{b^2 + c^2 + d^2}} k.$$

Exercise 6.166. *Let $p = a + bi + cj + dk$. Find all quaternions p satisfying the given equation:* (a) $\sin(p) = 0_{\mathbb{H}};$ (b) $\cos(p) = 0_{\mathbb{H}};$ (c) $\cos(p) = 1 - i;$ (d) $i \cos(p) j = k;$ (e) $\cos(p) = -\sin(p);$ (f) $\sin^2(p) = 2 \sin(p) - 1;$ (g) $2 \sin(p) \cos(p) = \sin(2p).$

6.4 Zeros

A standard result in real analysis says that the zeros (roots) of the real sine function are the integer multiples of π, and that the zeros of the real cosine function are the odd integer multiples of $\frac{\pi}{2}$. Analogous statements hold for the complex sine and cosine functions. It is clearly of interest to check whether similar conclusions hold in a quaternionic context. One way of answering these questions is by solving the equations $\sin(p) = 0_{\mathbb{H}}$ and $\cos(p) = 0_{\mathbb{H}}$ in the manner presented above. A different method involves recalling that a quaternion is equal to $0_{\mathbb{H}}$ if and only if its modulus is 0. This is a simple matter once the reader observes that $\sin(p) = 0_{\mathbb{H}}$ and

$\cos(p) = 0_{\mathbb{H}}$ are equivalent to solving the equations $|\sin(p)| = 0$ and $|\cos(p)| = 0$, respectively.

To make things understandable, we again rely on the expressions

$$\sin(p) = \sin(p_0)\cos(\mathbf{p}) + \cos(p_0)\sin(\mathbf{p}),$$

$$\cos(p) = \cos(p_0)\cos(\mathbf{p}) - \sin(p_0)\sin(\mathbf{p}).$$

Taking into account that $\operatorname{sgn}(\overline{\mathbf{p}}) = -\operatorname{sgn}(\mathbf{p})$, one has

$$|\sin(p)|^2 = \sin^2(p_0) + \sinh^2|\mathbf{p}|,$$

$$|\cos(p)|^2 = \cos^2(p_0) + \cosh^2|\mathbf{p}| - 1.$$

Since $\sin^2(p_0)$ (resp. $\cos^2(p_0)$) and $\sinh^2|\mathbf{p}|$ are both nonnegative real numbers, it is evident that the above equations are satisfied if and only if $\sin(p_0) = 0$ (resp. $\cos(p_0) = 0$) and $\sinh|\mathbf{p}| = 0$. As just noticed, $\sin(p_0) = 0$ when $p_0 = n\pi$, $n = 0, \pm1, \pm2, \ldots$ (resp. $\cos(p_0) = 0$ when $p_0 = \frac{(2n+1)\pi}{2}$, $n = 0, \pm1, \pm2, \ldots$), and $\sinh|\mathbf{p}| = 0$ only when $|\mathbf{p}| = 0$. Therefore the quaternion trigonometric functions have only the zeros known for the real functions, i.e.,

$$\sin(p) = 0_{\mathbb{H}} \quad \text{if and only if} \quad p = n\pi,$$

$$\cos(p) = 0_{\mathbb{H}} \quad \text{if and only if} \quad p = \left(n + \frac{1}{2}\right)\pi,$$

for $n = 0, \pm1, \pm2, \ldots$.

Exercise 6.167. *Let p be a nonnull pure quaternion. Prove the following relations:*
(a) $\sin(\mathbf{p}) = \operatorname{sgn}(\mathbf{p})\sinh|\mathbf{p}|$;   (b) $\cos(\mathbf{p}) = \cosh(|\mathbf{p}|)$.

Now we turn our attention to inequalities involving the quaternion trigonometric functions.

Example. Let p be a real quaternion such that $|\mathbf{p}| \leq \ln(1 + \sqrt{2})$. Prove that $|\sin(p)| \leq \sqrt{p_0^2 + 1}$.

Solution. Notice that $\sin^2(p_0) \leq p_0^2$, and $\sinh^2|\mathbf{p}| \leq 1$ for every $|\mathbf{p}| \in [0, \ln(1 + \sqrt{2})]$. From a previous representation, it follows that $|\sin(p)|^2 \leq p_0^2 + 1$.

Example. Let p be a real quaternion such that $|\mathbf{p}| \geq \ln(1 + \sqrt{2})$, and $p_0 \leq -\sqrt{6}$ or $p_0 \in [0, \sqrt{6}]$. Prove that $|\sin(p)| \geq \sqrt{1 + \left(p_0 - \frac{p_0^3}{6}\right)^2}$.

Solution. Note that $\sin^2(p_0) \geq \left(p_0 - \frac{p_0^3}{6}\right)^2$ when $p_0 \leq -\sqrt{6}$ or $p_0 \in [0, \sqrt{6}]$, and $\sinh^2(|\mathbf{p}|) \geq 1$ for $|\mathbf{p}| \leq \ln(1 + \sqrt{2})$. We find

$$|\sin(p)|^2 \geq 1 + \left(p_0 - \frac{p_0^3}{6}\right)^2.$$

Example. Let p be a real quaternion such that $p_0 \in [0, \sqrt{2}]$. Prove that $|\cos(p)| \geq 1 - \frac{p_0^2}{2}$.

Solution. Let $p_0 \in [0, \sqrt{2}]$. Then clearly $\cos^2(p_0) \geq \left(1 - \frac{p_0^2}{2}\right)^2$, and so

$$|\cos(p)|^2 \geq \left(1 - \frac{p_0^2}{2}\right)^2 + 1 - 1 = \left(1 - \frac{p_0^2}{2}\right)^2.$$

Further, we adopt the usual definitions of the other trigonometric functions:

6.5   Quaternion Tangent and Secant Functions

For $p \in \mathbb{H} \setminus \{(n + 1/2)\pi : n = 0, \pm 1, \pm 2, \ldots\}$, the functions $\tan(p)$ and $\sec(p)$, defined respectively by

$$\tan(p) := \frac{\sin(p)}{\cos(p)} \quad \text{and} \quad \sec(p) := \frac{1}{\cos(p)},$$

are called the quaternion tangent and secant functions. Based on the expressions for $\cos(p)$ and $\sin(p)$ considered above, we deduce the following identities:

$$\tan(p) = \frac{\sin(p_0)\cos(p_0)\left(|\cos(\mathbf{p})|^2 - |\sin(\mathbf{p})|^2\right)}{\cos^2(p_0) + \sinh^2|\mathbf{p}|}$$
$$+ \frac{\cos^2(p_0)\overline{\cos(\mathbf{p})\sin(\mathbf{p})} - \sin^2(p_0)\cos(\mathbf{p})\overline{\sin(\mathbf{p})}}{\cos^2(p_0) + \cosh^2|\mathbf{p}| - 1}$$

and

$$\sec(p) = \frac{\cos(p_0)\overline{\cos(\mathbf{p})} - \sin(p_0)\overline{\sin(\mathbf{p})}}{\cos^2(p_0) + \cosh^2|\mathbf{p}| - 1}.$$

Exercise 6.168. *Compute* $\tan(i)$.

Solution. $\tanh(1)$.

6.6   Quaternion Cotangent and Cosecant Functions

For $p \in \mathbb{H} \setminus \{n\pi : n = 0, \pm 1, \pm 2, \ldots\}$, the functions $\cot(p)$ and $\csc(p)$, defined respectively by

$$\cot(p) := \frac{\cos(p)}{\sin(p)} \quad \text{and} \quad \csc(p) := \frac{1}{\sin(p)},$$

are called the quaternion cotangent and cosecant functions. Based on the expressions for $\cos(p)$ and $\sin(p)$ considered above, we deduce the following identities:

$$\cot(p) = \frac{\sin(p_0)\cos(p_0)\left(|\cos(\mathbf{p})|^2 - |\sin(\mathbf{p})|^2\right)}{\cos^2(p_0) + \sinh^2|\mathbf{p}|}$$

$$+ \frac{\cos^2(p_0)\,\overline{\cos(\mathbf{p})\sin(\mathbf{p})} - \sin^2(p_0)\overline{\cos(\mathbf{p})\sin(\mathbf{p})}}{\cos^2(p_0) + \sinh^2|\mathbf{p}|}$$

and

$$\csc(p) = \frac{\sin(p_0)\overline{\cos(\mathbf{p})} + \cos(p_0)\overline{\sin(\mathbf{p})}}{\cos^2(p_0) + \sinh^2|\mathbf{p}|}.$$

Since the quaternion sine and cosine agree with the real and complex sine and cosine functions, the above-mentioned trigonometric functions also agree with their real and complex counterparts. These functions clearly share many other properties, but for the sake of brevity we will not review them here. Instead we suggest the reader to play with these functions until he is familiar with them.

Exercise 6.169. *Compute* $\cot(i)$.

Solution. $\coth(1)$.

Exercise 6.170. *Compute the value of the given quaternion trigonometric function:* (a) $\tan(i^k + j^k)$;   (b) $\cot(e^p)$;   (c) $\sec(p) - \tan(p)\cot(p)$;   (d) $\csc(\frac{1}{p})$.

Exercise 6.171. *Prove the following trigonometric identities:* (a) $\tan(-p) = -\tan(p)$;   (b) $\sec(-p) = \sec(p)$;   (c) $\cot(p) = -\cot(-p)$;   (d) $\csc(-p) = -\csc(p)$.

Exercise 6.172. *Is the quaternion tangent function periodic? Justify your answer.*

Solution. The tangent quaternion function is not periodic since the quaternion exponential function is not periodic.

Exercise 6.173. *Find all quaternions p such that:* (a) $|\tan(p)| = 1$; (b) $|\cot(p)| = 1$.

6.7 Advanced Practical Exercises

1. Compute (a) $\sin(i)$; (b) $\sin(j)$; (c) $\sin(k)$; (d) $\cos(i)$; (e) $\cos(j)$; (f) $\cos(k)$; (g) $\tan(i)$; (h) $\tan(j)$; (i) $\tan(k)$.
2. Compute (a) $\sin(1 + i - j + k)$; (b) $\cos(1 - k)$; (c) $\sin(1 - j)$; (d) $\sin(1 + i - j)$.
3. Compute (a) $\tan(i) + \cos(i + j) + \sin(k)$; (b) $\cos(i - k) + \sin(i)\cos(j) + \sin(i + j)$.
4. Solve the following equations (a) $\sin(p) = i$; (b) $\sin(p) = j$; (c) $\sin(p) = k$; (d) $\cos(p) = i$; (e) $\cos(p) = j$; (f) $\cos(p) = k$; (g) $\tan(p) = i$; (h) $\tan(p) = j$; (i) $\tan(p) = k$.
5. Solve the equations (a) $\sin(p) = 1 + i - j + k$; (b) $\cos(p) = 1 - i - j - k$.
6. Let $p \in \mathbb{H}$ such that $|\mathbf{p}| = 1$. Check the validity of the following relation: $\arg(p\mathbf{p}\overline{p}) = 2\arg(p)$.
7. Let p be pure quaternion such that $|p| = 1$. Prove that: (a) $\sin^2 p + \cos^2 p = 1$; (b) $\cos^2 p - \sin^2 p = \sinh(2|p|)$; (c) $\cos^2 p = \frac{1+\sinh(2|p|)}{2}$; (d) $\sin^2 p = \frac{1-\sinh(2|p|)}{2}$.
8. Let p be pure quaternion such that $p \neq 0_{\mathbb{H}}$. Prove that: (a) $\tan(p) = \operatorname{sgn}(p)\tanh(|p|)$; (b) $\cot(p) = \frac{1}{\operatorname{sgn}(p)}\coth(|p|)$.
9. Find a pure quaternion p such that (a) $\cos^2 p - \sin^2 p = 1 - i + j$; (b) $\cos^2 p - \sin^2 p = i + 2j$; (c) $\cos^2 p - \sin^2 p = 1_{\mathbb{H}}$; (d) $\cos^2 p - \sin^2 p = j + 10k$; (e) $\cos^2 p - \sin^2 p = 1 - 2j - 4k$; (f) $\cos^2 p - \sin^2 p = -3 + i + j + k$; (g) $\cos^2 p - \sin^2 p = i - j - 4k$; (h) $\cos^2 p - \sin^2 p = 8i + k$; (i) $\cos^2 p - \sin^2 p = 7 - i - j + 3k$; (j) $\cos^2 p - \sin^2 p = j - k$.
10. Find a pure quaternion p such that (a) $\cos^2 p = 1 - i + j$; (b) $\cos^2 p = i + 2j$; (c) $\cos^2 p = 1_{\mathbb{H}}$; (d) $\cos^2 p = j + 10k$; (e) $\cos^2 p = -1 - 2j - 4k$; (f) $\cos^2 p = -3 + i + j + k$; (g) $\cos^2 p = -i - j - 4k$; (h) $\cos^2 p = 8i + k$; (i) $\cos^2 p = 5 - j$; (j) $\cos^2 p = -j - k$.
11. Find a pure quaternion p such that (a) $\sin^2 p = 1 - i + j$; (b) $\sin^2 p = i + 2j$; (c) $\sin^2 p = 1_{\mathbb{H}}$; (d) $\sin^2 p = j + 10k$; (e) $\sin^2 p = 1 - 2j - 4k$; (f) $\sin^2 p = -3 + i + j + k$; (g) $\sin^2 p = i - j - 4k$; (h) $\sin^2 p = 8i + k$; (i) $\sin^2 p = 3 - i$; (j) $\sin^2 p = j - k$.
12. Let p be a pure quaternion. Prove that $\cos(2p) = 2\cos^2(p) - 1_{\mathbb{H}}$.
13. Let p be a pure quaternion. Prove each of the following limits: (a) $\lim_{p \to 0_{\mathbb{H}}} \cos p = \pm\frac{\sqrt{2}}{2}$; (b) $\lim_{p \to 0_{\mathbb{H}}} \sin p = \pm\frac{\sqrt{2}}{2}$.

Hint. Use the formulae $\cos^2 p = \frac{1+\sinh(2|p|)}{2}$ and $\sin^2 p = \frac{1-\sinh(2|p|)}{2}$.

14. Let p be a pure quaternion. Prove that $\lim_{p\to\infty} \cos(p) = \pm\infty$.

15. Let p be a pure quaternion. Prove that $\lim_{p\to\infty} \sin(p)$ does not exist.

16. Let p be a pure quaternion. Prove that $\lim_{p\to 0_{\mathbb{H}}} \left(\frac{\cos p}{p}\right)^2 = \infty$.

17. Let p be a pure quaternion. Prove that $\lim_{p\to 0_{\mathbb{H}}} \left(\frac{\sin p}{p}\right)^2 = \infty$.

Hyperbolic Functions **7**

After bringing together various results mentioned before, in this chapter we introduce the quaternion hyperbolic functions, whose study will require us to master a new situation. Since the quaternion exponential function agrees with the real and complex exponential function of real and complex arguments, it follows that the quaternion hyperbolic functions also agree with their usual counterparts for real and complex input. This allows us to discuss some important hyperbolic identities and the existence of infinitely many zeros of the quaternion sine and cosine hyperbolic functions, and to solve equations involving hyperbolic functions. A remarkable result of the theory exhibits the deep connection between the hyperbolic and trigonometric functions discussed in the previous chapter. We hope that material presented in this part will make this beautiful topic accessible to the reader.

The quaternion hyperbolic sine and cosine functions may be defined using the quaternion exponential function as follows:

7.1 Quaternion Sine and Cosine Hyperbolic Functions

The functions $\sinh(p)$ and $\cosh(p)$, defined by

$$\sinh(p) := \frac{e^p - e^{-p}}{2}, \qquad \cosh(p) := \frac{e^p + e^{-p}}{2},$$

respectively, are called the quaternion sine and cosine hyperbolic functions. These expressions make it clear that the quaternion hyperbolic functions are generalizations of the real and complex hyperbolic functions to quaternion values. Alternatively, the reader is doubtless familiar with the idea of introducing power series as another way of defining the quaternion sine and cosine hyperbolic functions and with the fact that both approaches yield the same quaternion function. The quaternion sine and cosine hyperbolic functions can be expressed via their Taylor series as follows:

J.P. Morais et al., *Real Quaternionic Calculus Handbook*,
DOI 10.1007/978-3-0348-0622-0_7, © Springer Basel 2014

$$\sinh(p) = \sum_{l=0}^{\infty} \frac{p^{2l+1}}{(2l+1)!} \quad \text{and} \quad \cosh(p) = \sum_{l=0}^{\infty} \frac{p^{2l}}{(2l)!}.$$

Example. Let $p = 1 - i - j$. Find $\sinh(p)$.

Solution. Direct computations show that

$$\sinh(1 - i - j) = \sinh(1)\cos(\sqrt{2}) - \frac{i+j}{2}\sin(\sqrt{2})\cosh(1).$$

Exercise 7.174. *Compute the value of the given quaternion hyperbolic function:* (a) $\cosh(i-j)$;   (b) $\cosh(i^j)$;   (c) $\sinh(i) + \cosh(j) + \mathrm{Ln}(k)$;   (d) $\sinh(\cosh(k)) + 2i^{\sinh(j)}$.

Solution. (a) $\cos\sqrt{2}$;   (b) $\cos 1$;   (c) $\cos 1 + i\sin 1 + \frac{\pi}{2}k$;   (d) $\sinh(\cos 1) + 2\cos(\frac{\pi}{2}\sin 1) + 2\sin(\frac{\pi}{2}\sin 1)k$.

Exercise 7.175. *Compute* (a) $\sum_{l=0}^{\infty} \frac{(i-j)^{2l+1}}{(2l+1)!}$;   (b) $\sum_{l=0}^{\infty} \frac{(i+j+k)^{2l}}{(2l)!}$

Solution. (a) $\sinh(i-j)$;   (b) $\cosh(i+j+k)$.

Exercise 7.176. *Compute the following quaternion power series:*

(a) $\left(\sum_{l=0}^{\infty} \frac{(1+i-j)^{2l+1}}{(2l+1)!}\right)\left(\sum_{l=0}^{\infty} \frac{(1+i-j)^{2l}}{(2l)!}\right)$;

(b) $\left(\sum_{l=0}^{\infty} \frac{(1+3i-j-k)^{2l+1}}{(2l+1)!}\right)\left(\sum_{l=0}^{\infty} \frac{(1+3i-j-k)^{2l}}{(2l)!}\right)$;

(c) $\left(\sum_{l=0}^{\infty} \frac{(1+i+j+k)^{2l+1}}{(2l+1)!}\right)^2$;

(d) $\left(\sum_{l=0}^{\infty} \frac{(1-i-j-k)^{2l+1}}{(2l+1)!}\right)^2$.

Solution. (a) $\frac{1}{2}\sinh(2+2i-2j)$;   (b) $\frac{1}{2}\sinh(2+6i-2j-2k)$;   (c) $\frac{1}{2}(\cosh(2+2i+2j+2k)-1)$;   (d) $\frac{1}{2}(\cosh(2-2i-2j-2k)-1)$.

7.2   Hyperbolic Identities

Similarly to their counterparts for real and complex input, the quaternion hyperbolic sine and cosine are, respectively, odd and even functions; that is, $-\sinh(p) = \sinh(-p)$, and $\cosh(p) = \cosh(-p)$. In particular, $\sinh(0_{\mathbb{H}}) = 0_{\mathbb{H}}$, while $\cosh(0_{\mathbb{H}}) = 1_{\mathbb{H}}$. In this sense our discussion can serve as a brief review of previously understood notions. A natural question may have occurred to some readers by now: Do familiar identities for real and complex hyperbolic functions still hold in the quaternionic context? A straightforward computation shows that for $\cosh^2(p) - \sinh^2(p) = 1_{\mathbb{H}}$ the answer is positive. The proof is left as an exercise. Setting $p = i - j$, it follows that

$$\cosh^2(i-j) - \sinh^2(i-j) = \cos^2(\sqrt{2}) - \frac{(i-j)^2}{2}\sin^2(\sqrt{2}) = 1_{\mathbb{H}}.$$

However, we must keep clearly in mind that in general the quaternion hyperbolic functions do not satisfy the sum and difference formulae $\sinh(p \pm q) = \sinh p \cosh q \pm \cosh p \sinh q$, and $\cosh(p \pm q) = \cosh p \cosh q \pm \sinh p \sinh q$, unless p and q commute. Here is a simple example: take $p = i - j$; then $\sinh(i - j) = \frac{(i-j)}{\sqrt{2}}\sin(\sqrt{2}) \neq (i-j)\cos(1)\sin(1) = \sinh(i)\cosh(j) - \cosh(i)\sinh(j)$.

Exercise 7.177. *Let $p \in \mathbb{H} \setminus \{0_{\mathbb{H}}\}$ and $\sinh(p) = i + j + k$. Find $\cosh^2(p)$.*

Solution. $\cosh^2(p) = -2$ for all nonzero $p \in \mathbb{H}$.

Exercise 7.178. *Let $p \in \mathbb{H} \setminus \{0_{\mathbb{H}}\}$ and $\cosh(p) = j + k$. Find $\sinh^2(p)$.*

Solution. $\sinh^2(p) = -3$ for all nonzero $p \in \mathbb{H}$.

Exercise 7.179. *Let $p \in \mathbb{H} \setminus \{0_{\mathbb{H}}\}$, $\sinh(p) = i - j + k$, and $\cosh(p) = 1 + k$. Find $\sinh(2p)$.*

Solution. $\sinh(2p) = -2 - 4j + 2k$ for all nonzero $p \in \mathbb{H}$.

Exercise 7.180. *Let $p \in \mathbb{H} \setminus \{0_{\mathbb{H}}\}$, $\sinh(2p) = 1 - i - j - k$, and $\cosh(p) = \frac{i+j}{2}$. Find $\sinh(p)$.*

Solution. $\sinh(p) = -1 - i$ for all nonzero $p \in \mathbb{H}$.

Exercise 7.181. *Let $p \in \mathbb{H} \setminus \{0_{\mathbb{H}}\}$, $\sinh(2p) = i + j$, and $\sinh(p) = k$. Find $\cosh(p)$.*

Solution. $\cosh(p) = \frac{1}{2}(i - j)$ for all nonzero $p \in \mathbb{H}$.

Exercise 7.182. *Let $p \in \mathbb{H} \setminus \{0_{\mathbb{H}}\}$ and $\sinh(p) = i + j$. Find $\cosh(2p)$.*

Solution. $\cosh(2p) = -3$ for all nonzero $p \in \mathbb{H}$.

Exercise 7.183. *Under what circumstances do the quaternion hyperbolic functions satisfy the sum and difference formulae?*

Solution. For all quaternions p and q for which $pq = qp$.

Exercise 7.184. *Prove the identities:* (a) $\sinh(2p) = 2\sinh(p)\cosh(p)$; (b) $\cosh(2p) = 2\sinh^2(p) + 1$; (c) $\sinh(3p) = -\sinh(p) + 4\sinh(p)\cosh^2(p)$; (d) $\sinh(4p) = 8\sinh^3(p)\cosh(p) + 4\sinh(p)\cosh(p)$; (e) $\cosh(3p) = 4\cosh^3(p) - 3\cosh(p)$; (f) $\cosh(4p) = 8\sinh^2(p)\cosh^2(p) + 1$.

7.3   Equations with Hyperbolic Functions

The above considerations enable us to gain insight into equations involving the aforementioned hyperbolic functions. For the convenience of the reader, we note here that for a given quaternion w equations of the form $\sinh(p) = w$ or $\cosh(p) = w$ always either have no solutions, or have finitely many or infinitely many solutions; this has to do with the possible many-valuedness of these functions. To exemplify this, let $p = cj$ be a solution to the equation $\cosh(p) = 1_{\mathbb{H}}$. A direct computation shows that $\cos(|c|) = 1$, and we see from this that $p = \pm 2\pi n\,j,\, n \in \mathbb{Z}$. We can verify that each of these values of p satisfies the equation $\cosh(p) = 1_{\mathbb{H}}$.

Example. Let $p = a + bi + cj + dk$, where $a, b, c, d \in \mathbb{H}$. Solve the equation $\sinh(p) = \frac{\sqrt{3}}{3}$.

Solution. We first compute $\sinh(p)$ explicitly:

$$\sinh(p) = \sinh(a) \cos\left(\sqrt{b^2 + c^2 + d^2}\right)$$

$$+ \cosh(a) \frac{\sin\left(\sqrt{b^2 + c^2 + d^2}\right)}{\sqrt{b^2 + c^2 + d^2}} (bi + cj + dk).$$

We obtain the following system for the parameters a, b, c, d:

$$\begin{cases} \sinh(a) \cos\left(\sqrt{b^2 + c^2 + d^2}\right) = \dfrac{\sqrt{3}}{3}, \\[2ex] b\cosh(a) \dfrac{\sin\left(\sqrt{b^2 + c^2 + d^2}\right)}{\sqrt{b^2 + c^2 + d^2}} = 0, \\[2ex] c\cosh(a) \dfrac{\sin\left(\sqrt{b^2 + c^2 + d^2}\right)}{\sqrt{b^2 + c^2 + d^2}} = 0, \\[2ex] d\cosh(a) \dfrac{\sin\left(\sqrt{b^2 + c^2 + d^2}\right)}{\sqrt{b^2 + c^2 + d^2}} = 0. \end{cases}$$

This system yields that $b = c = d = 0$, and a is such that $\cosh(a) = \frac{\sqrt{3}}{3}$. Hence $a = \operatorname{arccosh}(\frac{\sqrt{3}}{3})$. We conclude that $p = \operatorname{arccosh}(\frac{\sqrt{3}}{3})$.

Exercise 7.185. *Let* $p = a + bi + cj + dk$. *Find* $\cosh(p)$.

Solution. The value is computed as

$$\cosh(a)\cos\left(\sqrt{b^2+c^2+d^2}\right) + \sinh(a)\frac{\sin\left(\sqrt{b^2+c^2+d^2}\right)}{\sqrt{b^2+c^2+d^2}}(bi+cj+dk).$$

Exercise 7.186. *Let $p = a + bi + cj + dk$. Find all quaternions p satisfying the given equation:* (a) $\sinh(p) = 0_{\mathbb{H}}$; (b) $\cosh(p) = 0_{\mathbb{H}}$; (c) $\sinh(p) = -1_{\mathbb{H}}$; (d) $\sinh(p) - \cosh(p) = i - j + 2k$; (e) $\cosh(p) = e^{-p}$.

7.4   Relation to Quaternion Sine and Cosine Functions

We are now in a position to gather and set some light on the diverse notions concerning the quaternion trigonometric and hyperbolic functions from a unifying point of view. The motivation for this exposition comes in part from the simple and direct relation between these functions. Our intention is to familiarize the reader with these relations. By definition,

$$e^{\pm\mathbf{p}} = \cos|\mathbf{p}| \pm \mathrm{sgn}(\mathbf{p})\sin|\mathbf{p}|,$$

so we see that quaternion trigonometric and hyperbolic functions are simply related:

$$\cosh(\mathbf{p}) = \cos|\mathbf{p}|,$$

and in a similar manner

$$\sinh(\mathbf{p}) = \mathrm{sgn}(\mathbf{p})\sin|\mathbf{p}|.$$

Using the sum and difference formulae, a straightforward calculation exhibits a noteworthy connection between the trigonometric and the hyperbolic functions, namely,

$$\sinh(p) = \sinh(p_0)\cos|\mathbf{p}| + \cosh(p_0)\sinh(\mathbf{p}),$$

$$\cosh(p) = \cosh(p_0)\cos|\mathbf{p}| + \sinh(p_0)\sinh(\mathbf{p}).$$

Exercise 7.187. *Prove the following identities:* (a) $\cosh(p) = \cos(\mathrm{sgn}(\mathbf{p}\,p))$; (b) $\sinh(p) = \mathrm{sgn}(\mathbf{p})\sin(\mathrm{sgn}(\mathbf{p}\,p))$; (c) $\cosh(\mathbf{p}) = \cos|\mathbf{p}|$; (d) $\sinh(\mathbf{p}) = -\mathrm{sgn}(\mathbf{p})\sin|\mathbf{p}|$.

7.5   Zeros

In what follows, we shall be interested in determining the zeros of the quaternion sine and cosine hyperbolic functions, i.e., the solutions of the equations $\sinh(p) = 0_{\mathbb{H}}$ and $\cosh(p) = 0_{\mathbb{H}}$. In order to fully understand the meaning of these expressions, we must keep in mind that the functions $\sinh(p)$ and $\cosh(p)$ are not periodic and have infinitely many zeros; we must further have a clear agreement on how both notions should be mastered. The zeros of the function $\sinh(p)$ are the solutions for p of the equation $e^p = e^{-p}$, which can be rewritten as $e^{2p} = 1_{\mathbb{H}}$. Since $e^{2p} = 1_{\mathbb{H}}$ only when $|\mathrm{Vec}(2p)|$ is an integer multiple of $2\pi\,\mathrm{sgn}(\mathbf{p})$, we conclude that the zeros of $\sinh(p)$ are the numbers $n\pi\,\mathrm{sgn}(\mathbf{p})$, $n = 0, \pm 1, \pm 2, \ldots$. Similarly, the zeros of $\cosh(p)$ are the solutions for p of the equation $e^p = -e^{-p}$, which can be rewritten as $e^{2p} = -1_{\mathbb{H}}$, or as $e^{2p - \mathrm{sgn}(\mathbf{p})\pi} = 1_{\mathbb{H}}$ (since $e^{\mathrm{sgn}(\mathbf{p})\pi} = -1_{\mathbb{H}}$). It follows that the zeros of $\cosh(p)$ are the numbers $\left(n + \frac{1}{2}\right)\pi\,\mathrm{sgn}(\mathbf{p})$ with $n = 0, \pm 1, \pm 2, \ldots$.

We proceed to study some inequalities involving the quaternion sine and cosine hyperbolic functions.

Example. Let p be a real quaternion. Prove that $|\sinh(p)| \leq 2\cosh(p_0)$.

Solution. We note that $|e^p| \leq 2e^{p_0}$. In particular, $|e^{-p}| \leq 2e^{-p_0}$. It follows that $|\sinh(p)| \leq 2\cosh(p_0)$.

Exercise 7.188. *Let p be a quaternion. Prove that $|\cosh(p)| \leq 2\cosh(p_0)$.*

Exercise 7.189. *Let $p = 1 + i + j + k$. Prove the following inequalities:* (a) $|\sinh(p)|^n \leq 2^n \cosh^n(1)$ *for all $n \in \mathbb{N}$;* (b) $|\cosh(p)|^n \leq 2^n \cosh^n(1)$ *for all $n \in \mathbb{N}$.*

Exercise 7.190. *Let $p \in \mathbb{H}$. Prove that*
(a) $|\sinh(2p)| \leq 8\cosh^2(p_0)$;
(b) $|\cosh(2p)| \leq 8\cosh^2(p_0) + 1$;
(c) $|\sinh(3p)| \leq 32\cosh^3(p_0) + 2\cosh(p_0)$;
(d) $|\cosh(3p)| \leq 32\cosh^3(p_0) + 6\cosh(p_0)$;
(e) $|\sinh(4p)| \leq 16\cosh^2(p_0)(8\cosh^2(p_0) + 1)$;
(f) $|\cosh(4p)| \leq 128\cosh^4(p_0) + 1$.

Analogously to quaternion trigonometric functions, we next define the quaternion tangent, cotangent, secant, and cosecant hyperbolic functions using the quaternion hyperbolic sine and cosine.

7.6   Quaternion Tangent and Secant Functions

The functions $\tanh(p)$ and $\mathrm{sech}(p)$, defined respectively by

$$\tanh(p) := \frac{\sinh(p)}{\cosh(p)} \quad \text{and} \quad \mathrm{sech}(p) := \frac{1}{\cosh(p)},$$

for $p \in \mathbb{H} \setminus \{(n + \frac{1}{2})\pi\,\mathrm{sgn}(\mathbf{p}) : n = 0, \pm 1, \pm 2, \ldots\}$, are called the quaternion tangent and secant hyperbolic functions.

7.7 Quaternion Cotangent and Cosecant Hyperbolic Functions

The functions $\coth(p)$ and $\operatorname{csch}(p)$, defined respectively by

$$\coth(p) := \frac{\cosh(p)}{\sinh(p)} \quad \text{and} \quad \operatorname{csch}(p) := \frac{1}{\sinh(p)},$$

for $p \in \mathbb{H} \setminus \{n\pi \operatorname{sgn}(\mathbf{p}) : n = 0, \pm 1, \pm 2, \ldots\}$, are called the quaternion cotangent and cosecant functions.

Since the quaternion sine and cosine hyperbolic functions agree with the real and complex sine and cosine hyperbolic functions, these functions also agree with their counterparts for real and complex input.

Exercise 7.191. *Prove the following identities:* (a) $\tanh(-p) = -\tanh(p)$;   (b) $\coth(-p) = -\coth(p)$;   (c) $\operatorname{sech}(-p) = \operatorname{sech}(p)$;   (d) $\operatorname{csch}(-p) = -\operatorname{csch}(p)$.

Exercise 7.192. *Compute the value of the given quaternion hyperbolic function:* (a) $\coth(\frac{\pi}{2\sqrt{2}}(i+j))$;   (b) $\tanh(-i+k)$;   (c) $e^{\operatorname{csch}(1/p)}$.

Exercise 7.193. *Prove the following hyperbolic identities:* (a) $\tanh^2(p) = 1_{\mathbb{H}} - \operatorname{sech}^2(p)$;   (b) $\coth^2(p) = 1_{\mathbb{H}} + \operatorname{csch}^2(p)$.

Exercise 7.194. *Verify the validity of the following identities:* (a) $\sin(n\pi + \mathbf{p}) = \operatorname{sgn}(\mathbf{p})(-1)^n \sinh(\mathbf{p})$;   (b) $\cos(n\pi + \mathbf{p}) = (-1)^n \cosh(\operatorname{sgn}(\mathbf{p}\,p))$ *for* $n = 0, \pm 1, \pm 2, \ldots$.

7.8 Advanced Practical Exercises

1. Let $p = 1 + i - j - k$. Compute (a) $\sinh(i)$;   (b) $\sinh(j)$;   (c) $\sinh(k)$;   (d) $\cosh(i)$;   (e) $\cosh(j)$;   (f) $\cosh(k)$;   (g) $\tanh(i)$;   (h) $\tanh(j)$;   (i) $\tanh(k)$;   (j) $\sinh(p)$;   (k) $\cosh(p)$.
2. Let $p = i + j + k$. Find (a) $\sinh(p)$;   (b) $\cosh(p)$.
3. Let $p = 1 + i + j + k$. Find (a) $\sinh(p)$;   (b) $\cosh(p)$.
4. Solve the equations (a) $\sinh(p) = i$;   (b) $\sinh(p) = j$;   (c) $\sinh(p) = k$;   (d) $\cosh(p) = i$;   (e) $\cosh(p) = j$;   (f) $\cosh(p) = k$.
5. Solve the equations (a) $\sinh(p) = 1 + i + j + k$;   (b) $\cosh(p) = i - j + 2k$.
6. Compute (a) $pq - (p, q) + \sinh(r) - \overline{r}q + q \cdot r$;   (b) $\arg(p)q + e^r + \ln(q) + i^r - \cosh(r)$, where $p = 1 + i - j - k$, $q = 1 + i$, and $r = k$.
7. Compute (a) $i^{\sinh(k)}$;   (b) $i^{\sinh(j)}$;   (c) $j^{\sinh(i)}$;   (d) $j^{\sinh(k)}$;   (e) $k^{\sinh(i)}$;   (f) $k^{\sinh(j)}$.
8. Let p be nonnull pure quaternion. Prove that (a) $\sinh(p) = \frac{p}{|p|}\sin|p|$;   (b) $\cosh(p) = \cos|p|$;   (c) $\tanh(p) = \frac{p}{|p|}\tan|p|$;   (d) $\coth(p) = \cos|p|$.

9. Let p be a nonnull pure quaternion. Prove that $\lim_{p\to 0_{\mathbb{H}}} \sinh(p) = 0_{\mathbb{H}}$.

10. Let p be a nonnull pure quaternion. Prove that $\lim_{p\to 0_{\mathbb{H}}} \cosh(p) = 1_{\mathbb{H}}$.

11. Let p be a nonnull pure quaternion. Prove that $\cosh^2(p) - \sinh^2(p) = 1_{\mathbb{H}}$.

12. Let p be a nonnull pure quaternion. Prove that $\cosh^2 p + \sinh^2 p = \cos(2|p|)$.

Inverse Hyperbolic and Trigonometric Functions

8

The main focus of this chapter is to study the inverses of the quaternion trigonometric and hyperbolic functions, and their properties. Since the quaternion trigonometric and hyperbolic functions are defined in terms of the quaternion exponential function e^p, it can be shown that their inverses are necessarily multi-valued and can be computed via the quaternion natural logarithm function $\ln(p)$. The remarkable facts we shall see here attest the great interest of these functions in mathematics. Proofs of the most known facts are ommited.

In the sequel we shall recapitulate that for a given quaternion w equations of the form $\sinh(p) = w$ and $\cosh(p) = w$ either have no solutions, or have finitely many or infinitely many solutions. The procedure outlined a few pages back to find an explicit formula for w is often tedious and lengthy if all the details are taken into account, and so it is usually adequate to modify it. We summarize this discussion in the following definition.

8.1 Quaternion Inverse Hyperbolic Sine and Cosine Functions

The functions $\sinh^{-1}(p)$ and $\cosh^{-1}(p)$, defined respectively by

$$\sinh^{-1}(p) := \ln\left(p + \sqrt{p^2 + 1_{\mathbb{H}}}\right) \quad \text{and} \quad \cosh^{-1}(p) := \ln\left(p + \sqrt{p^2 - 1_{\mathbb{H}}}\right)$$

are called the quaternion inverse hyperbolic sine and cosine. Notice that each value of $w = \sinh^{-1}(p)$ satisfies the equation $\sinh(w) = p$, and, similarly, each value of $w = \cosh^{-1}(p)$ satisfies the equation $\cosh(w) = p$. Since the quaternion logarithm function agrees with the real and complex logarithm functions of real and complex arguments, these functions also agree with their counterparts for real and complex argument. Here we also call the inverse hyperbolic sine (resp. cosine) the arcsinh (resp. arccosh), and we will denote it by $\operatorname{arcsinh}(p)$ (resp. $\operatorname{arccosh}(p)$), as we shall always do in the sequel whenever we speak of inverse hyperbolic functions. These functions have two sources of multi-valuedness, one due to the quaternion natural

J.P. Morais et al., *Real Quaternionic Calculus Handbook*,
DOI 10.1007/978-3-0348-0622-0_8, © Springer Basel 2014

logarithm function $\ln(p)$, the other due to the quaternion power functions involved. It is evident that the quaternion inverse hyperbolic sine and cosine can be made single-valued by specifying a single value of the quaternion logarithm and a single value of the functions $(p^2+1_{\mathbb{H}})^{1/2}$ and $(p^2-1_{\mathbb{H}})^{1/2}$, respectively. We see at the same time that a branch of a quaternion inverse hyperbolic function may be obtained by choosing a branch of the quaternion logarithm and a branch of a quaternion power function. In spite of its local multi-valuedness, the $\ln(p)$ function has an infinite number of branches, hence so do the quaternion inverse hyperbolic sine and cosine. In addition, the quaternion inverse hyperbolic functions have branch point-solutions to the equations $p^2 \pm 1_{\mathbb{H}} = 0_{\mathbb{H}}$, because the functions $(p^2+1_{\mathbb{H}})^{1/2}$ and $(p^2-1_{\mathbb{H}})^{1/2}$ have no solutions of $(p^2 + 1_{\mathbb{H}})^{1/2} = 0_{\mathbb{H}}$ and $(p^2 - 1_{\mathbb{H}})^{1/2} = 0_{\mathbb{H}}$, respectively. Because this is an introductory text, we will not discuss this topic further.

Exercise 8.195. *Let $p = a + bi + cj + dk$, where $a,b,c,d \in \mathbb{H}$. Derive the formulae for* $\operatorname{arcsinh}(p)$ *and* $\operatorname{arccosh}(p)$.

Exercise 8.196. *Compute the value of the inverse hyperbolic functions* $\operatorname{arcsinh}(p)$ *and* $\operatorname{arccosh}(p)$ *at the given point:* (a) $p = j$; (b) $p = j^k - i$; (c) $p = i - j + k$; (d) $p = i - j^k$.

Solution. (a) $j\frac{\pi}{2}$, $i\frac{\pi}{2}$; (b) 0, $i\frac{\pi}{2}$; (c) $\log_e(5 + 2\sqrt{2}) + \frac{(1+\sqrt{2})i-j+k}{5+2\sqrt{2}}\frac{\pi}{2}$; $\log_e \sqrt{11} + \frac{3i-j+k}{\sqrt{11}}\frac{\pi}{2}$; (d) 0, $\frac{i\pi}{2}$.

Exercise 8.197. *Prove the identities:* (a) $\operatorname{arcsinh}(-p) = -\operatorname{arcsinh}(p)$; (b) $\operatorname{arccosh}(-p) = \operatorname{arccosh}(p)$; (c) $\operatorname{arccosh}(2p^2 - 1_{\mathbb{H}}) = 2\operatorname{arccosh}(p)$; (d) $2\operatorname{arcsinh}(p) = \operatorname{arccosh}(2p^2 + 1_{\mathbb{H}})$; (e) $\operatorname{arccosh}(p) = 2\operatorname{arcsinh}\sqrt{\frac{p-1_{\mathbb{H}}}{2}}$; (f) $\operatorname{arccosh}(p) = 2\operatorname{arccosh}\sqrt{\frac{p+1_{\mathbb{H}}}{2}}$; (g) $\sinh(\operatorname{arccosh}(p)) = \sqrt{p^2 - 1_{\mathbb{H}}}$; (h) $\cosh(\operatorname{arcsinh}(p)) = \sqrt{p^2 + 1_{\mathbb{H}}}$; (i) $\cosh(2\operatorname{arcsinh}p) = 2p^2 + 1_{\mathbb{H}}$; (j) $\sinh(2\operatorname{arccosh}(p)) = 2p\sqrt{p^2 - 1_{\mathbb{H}}}$.

Exercise 8.198. *Prove the identities:* (a) $\tanh(\operatorname{arccosh}(p)) = \sqrt{p^2 - 1}\,p^{-1}$; (b) $\coth(\operatorname{arccosh}(p)) = p\left(\sqrt{p^2 - 1}\right)^{-1}$; (c) $\tanh(\operatorname{arcsinh}(p)) = p\left(\sqrt{p^2 + 1}\right)^{-1}$; (d) $\coth(\operatorname{arcsinh}(p)) = \sqrt{p^2 + 1}\,p^{-1}$.

Exercise 8.199. *Prove the inequalities:* (a) $|\operatorname{arcsinh}(1 + i + j + k)| \leq \sqrt{29} + \pi$; (b) $|\operatorname{arccosh}(1 + i + j + k)| \leq \sqrt{29} + \pi$.

Exercise 8.200. *Let p be a real quaternion such that $|\mathbf{p}| \neq 0$, and*

$$1 + 2p_0\sqrt{4(p_0)^2|\mathbf{p}|^2 - 3\left(p_0^2 + 1 - |\mathbf{p}|^2\right)^2} \neq 0.$$

Prove that

$$\operatorname{arcsinh}(p) = \log_e |p_0 + C|$$

$$+ \operatorname{sgn}\left(\mathbf{p} + \operatorname{sgn}(B)\sqrt{A^2 - C^2}\right)\left(\arccos \frac{p_0 + C}{|p_0 + C + \mathbf{p}\operatorname{sgn}(B)\sqrt{A^2 - C^2}|} + 2n\pi\right),$$

for $n \in \mathbb{Z}$, where

$$A := \sqrt{(p_0^2 + 1 - |\mathbf{p}|^2)^2 + 4p_0^2|\mathbf{p}|^2}, \quad B := 2p_0\mathbf{p}, \quad C := p_0^2 + 1 - |\mathbf{p}|^2.$$

We proceed to study certain inequalities involving the quaternion inverse hyperbolic sine and cosine functions.

Example. Let p be a real quaternion such that $|p|^4 - |p|^2 + 2p_0^2 \geq 0$. Prove that
$$|\operatorname{arcsinh}(p)| \leq \sqrt{1 + 2(|p|^4 - |p|^2 + 2p_0^2)} + \pi.$$

Solution. To begin with, we find the following representation for $p + \sqrt{p^2 + 1_{\mathbb{H}}}$:

$$p + \sqrt{(p_0^2+1 - |\mathbf{p}|^2)^2 + 4p_0^2|\mathbf{p}|^2}\left(\cos\left|\frac{1}{2}\arccos\frac{p_0^2 + 1 - |\mathbf{p}|^2}{(p_0^2 + 1 - |\mathbf{p}|^2)^2 + 4p_0^2|\mathbf{p}|^2}\right|\right.$$

$$+ \operatorname{sgn}(\mathbf{p})\sin\left|\frac{1}{2}\arccos\frac{p_0^2 + 1 - |\mathbf{p}|^2}{(p_0^2 + 1 - |\mathbf{p}|^2)^2 + 4p_0^2|\mathbf{p}|^2}\right|\Bigg). \tag{8.1}$$

With A denoting the term $\sqrt{(p_0^2 + 1 - |\mathbf{p}|^2)^2 + 4p_0^2|\mathbf{p}|^2}$, we have, on account of (8.1):

$$|p + \sqrt{p^2 + 1_{\mathbb{H}}}|^2 = \left(p_0 + A\cos\left|\frac{1}{2}\arccos\frac{p_0^2 + 1 - |\mathbf{p}|^2}{(p_0^2 + 1 - |\mathbf{p}|^2)^2 + 4p_0^2|\mathbf{p}|^2}\right|\right)^2$$

$$+ \left(p_1 + A\frac{p_1}{\sqrt{p_1^2 + p_2^2 + p_3^2}}\sin\left|\frac{1}{2}\arccos\frac{p_0^2 + 1 - |\mathbf{p}|^2}{(p_0^2 + 1 - |\mathbf{p}|^2)^2 + 4p_0^2|\mathbf{p}|^2}\right|\right)^2$$

$$+ \left(p_2 + A\frac{p_2}{\sqrt{p_1^2 + p_2^2 + p_3^2}}\sin\left|\frac{1}{2}\arccos\frac{p_0^2 + 1 - |\mathbf{p}|^2}{(p_0^2 + 1 - |\mathbf{p}|^2)^2 + 4p_0^2|\mathbf{p}|^2}\right|\right)^2$$

$$+ \left(p_3 + A\frac{p_3}{\sqrt{p_1^2 + p_2^2 + p_3^2}}\sin\left|\frac{1}{2}\arccos\frac{p_0^2 + 1 - |\mathbf{p}|^2}{(p_0^2 + 1 - |\mathbf{p}|^2)^2 + 4p_0^2|\mathbf{p}|^2}\right|\right)^2.$$

Now we make use of the standard inequality $(a + b)^2 \leq 2(a^2 + b^2)$. Hence we obtain

$$
|p + \sqrt{p^2 + 1}_{\mathbb{H}}|^2 \leq 2\left[p_0^2 + \left(A\cos\left| \frac{1}{2}\arccos\frac{p_0^2 + 1 - |\mathbf{p}|^2}{(p_0^2 + 1 - |\mathbf{p}|^2)^2 + 4p_0^2|\mathbf{p}|^2} \right| \right)^2 \right.
$$

$$
+ p_1^2 + \left(A\frac{p_1}{\sqrt{p_1^2 + p_2^2 + p_3^2}}\sin\left| \frac{1}{2}\arccos\frac{p_0^2 + 1 - |\mathbf{p}|^2}{(p_0^2 + 1 - |\mathbf{p}|^2)^2 + 4p_0^2|\mathbf{p}|^2} \right| \right)^2
$$

$$
+ p_2^2 + \left(A\frac{p_2}{\sqrt{p_1^2 + p_2^2 + p_3^2}}\sin\left| \frac{1}{2}\arccos\frac{p_0^2 + 1 - |\mathbf{p}|^2}{(p_0^2 + 1 - |\mathbf{p}|^2)^2 + 4p_0^2|\mathbf{p}|^2} \right| \right)^2
$$

$$
\left. + p_3^2 + \left(A\frac{p_3}{\sqrt{p_1^2 + p_2^2 + p_3^2}}\sin\left| \frac{1}{2}\arccos\frac{p_0^2 + 1 - |\mathbf{p}|^2}{(p_0^2 + 1 - |\mathbf{p}|^2)^2 + 4p_0^2|\mathbf{p}|^2} \right| \right)^2 \right]
$$

$$
= 2\left(|p|^4 - |p|^2 + 2p_0^2 + 1 \right),
$$

that is,

$$
|p + \sqrt{p^2 + 1}_{\mathbb{H}}|^2 \leq 1 + 1 + 2\left(|p|^4 - |p|^2 + 2p_0^2 \right).
$$

Furthermore, using the fact that $\sqrt{a + b} \leq \sqrt{a} + \sqrt{b}$ for $a \geq 0$ and $b \geq 0$, the following further inequality is now immediate:

$$
\log_e |p + \sqrt{p^2 + 1}_{\mathbb{H}}| \leq \log_e\left(1 + \sqrt{1 + 2(|p|^4 - |p|^2 + 2p_0^2)} \right).
$$

We may now use the fact that $\ln(1 + x) \leq x$ for $x \geq 0$ to obtain

$$
\log_e |p + \sqrt{p^2 + 1}| \leq \sqrt{1 + 2\left(|p|^4 - |p|^2 + 2p_0^2 \right)},
$$

whence

$$
|\mathrm{Ln}(p + \sqrt{p^2 + 1})| \leq \sqrt{1 + 2\left(|p|^4 - |p|^2 + 2p_0^2 \right)} + \pi.
$$

We have thus obtained the inequality

$$
|\mathrm{arcsinh}(p)| \leq \sqrt{1 + 2\left(|p|^4 - |p|^2 + 2p_0^2 \right)} + \pi.
$$

Exercise 8.201. *Let p be a real quaternion such that $|p|^4 - |p|^2 + 2p_0^2 \geq 0$. Prove that*

$$|\operatorname{arccosh}(p)| \leq \sqrt{1 + 2(|p|^4 - |p|^2 + 2p_0^2)} + \pi.$$

Exercise 8.202. *Let p be a real quaternion such that $|p|^4 - |p|^2 + 2p_0^2 \geq 0$. Prove the following inequalities:*

(a) $|\operatorname{arcsinh}(p)| \leq \sum_{k=1}^{2n-1} (-1)^{k+1} \dfrac{\left(\sqrt{1+2(|p|^4-|p|^2+2p_0^2)}\right)^k}{k} + \pi, \quad n \in \mathbb{N};$

(b) $|\operatorname{arccosh}(p)| \leq \sum_{k=1}^{2n-1} (-1)^{k+1} \dfrac{\left(\sqrt{1+2(|p|^4-|p|^2+2p_0^2)}\right)^k}{k} + \pi, \quad n \in \mathbb{N}.$

All this being established, we can now introduce the inverse tangent.

8.2   Quaternion Inverse Hyperbolic Tangent Function

The function $\tanh^{-1}(p)$, defined by

$$\tanh^{-1}(p) := \frac{\ln(1+p) - \ln(1-p)}{2},$$

is called the quaternion inverse hyperbolic tangent. In particular, if $w = \tanh^{-1}(p)$, then $\tanh(w) = p$. This function, too, agrees with its real and complex counterparts. We shall frequently make use of another notation for the inverse hyperbolic tangent, namely, $\operatorname{arctanh}(p)$. We further assume the reader to be familiar with the fact that the inverse tangent is a multi-valued function, since it is defined in terms of the quaternion logarithm function $\ln(p)$. In order to supply additional information, we reiterate that the inverse hyperbolic tangent can be made single-valued by just specifying a single value of the quaternion logarithm.

Similar to the real and complex inverse hyperbolic functions, we define the quaternion inverse hyperbolic cotangent, secant, and cosecant functions using the quaternion inverse hyperbolic sine and cosine:

$$\operatorname{arccoth}(p) := \frac{\operatorname{arcsinh}(p)}{\operatorname{arccosh}(p)}, \quad \operatorname{arcsech}(p) := \frac{1}{\operatorname{arccosh}(p)}, \quad \operatorname{arccsch}(p) := \frac{1}{\operatorname{arcsinh}(p)}.$$

Exercise 8.203. *Compute the value of the function* $\operatorname{arctanh}(p)$ *at the given point:*
(a) $p = \ln(1_{\mathbb{H}});$ (b) $p = -i + j + k;$ (c) $p = i - j^k;$ (d) $p = \operatorname{arcsinh}(-j).$

Solution. (a) $0_{\mathbb{H}};$ (b) $\frac{2\pi\sqrt{3}}{9}(-i + j + k);$ (c) $0_{\mathbb{H}};$ (d) $1 + \frac{2}{\pi} \log_e \sqrt{\frac{\pi^2}{2} + 1}\, j.$

Exercise 8.204. *Prove the following identity:* $\operatorname{arctanh}(-p) = -\operatorname{arctanh}(p).$

In what follows, we shall study the inverse trignometric sine, cosine and tangent functions, which we now introduce.

8.3   Quaternion Inverse Trignometric Sine and Cosine Functions

The functions $\sin^{-1}(p)$ and $\cos^{-1}(p)$, defined respectively by

$$\sin^{-1}(p) := \operatorname{sgn}(\mathbf{p})\operatorname{arcsinh}\left(p\operatorname{sgn}(\mathbf{p})\right),$$

$$\cos^{-1}(p) := \operatorname{sgn}(\mathbf{p})\operatorname{arccosh}(p),$$

are called the quaternion inverse sine and cosine. Each value of $w = \sin^{-1}(p)$ satisfies the equation $\sin(w) = p$, and, similarly, each value of $w = \cos^{-1}(p)$ satisfies the equation $\cos(w) = p$. We also call the inverse sine (resp. cosine) the arcsine (resp. arccosine), and we will denote it by $\arcsin(p)$ (resp. $\arccos(p)$). As the above discussion shows, the inverse sine and cosine are necessarily multi-valued functions.

8.4   Quaternion Inverse Trignometric Tangent Function

The multi-valued function $\tan^{-1}(p)$, defined by

$$\tan^{-1}(p) := \operatorname{sgn}(\mathbf{p})\operatorname{arctanh}\left(p\operatorname{sgn}(\mathbf{p})\right),$$

is called the quaternion inverse tangent. Each value of $w = \tan^{-1}(p)$ satisfies the equation $\tan(w) = p$. The inverse tangent is often called arctangent, and will be denoted from now on by $\arctan(p)$.

Analogous to the real and complex inverse trigonometric functions, we define the quaternion inverse trigonometric cotangent, secant, and cosecant functions using the quaternion inverse trigonometric sine and cosine:

$$\operatorname{arccot}(p) := \frac{\arcsin(p)}{\arccos(p)}, \qquad \operatorname{arcsec}(p) := \frac{1}{\arccos(p)}, \qquad \operatorname{arccsc}(p) := \frac{1}{\arcsin(p)}.$$

Exercise 8.205. *Compute the value of the functions* $\arcsin(p)$, $\arccos(p)$ *and* $\arctan(p)$ *at the given point:* (a) $p = (-1 + 9k)e^{2i}$; (b) $p = \ln(\cos(i))$; (c) $p = \cosh(e) - e^k \sinh(1)$.

Exercise 8.206. *Prove the identities:* (a) $\arcsin(-p) = -\arcsin(p)$; (b) $\arccos(-p) = \arccos(p)$; (c) $\arctan(-p) = -\arctan(p)$.

8.5 Advanced Practical Exercises

1. Compute (a) $\operatorname{arcsinh}(1+i)$; (b) $\operatorname{arcsinh}(i)$; (c) $\operatorname{arcsinh}(j)$; (d) $\operatorname{arcsinh}(k)$;
   (e) $\operatorname{arccosh}(i)$; (f) $\operatorname{arccosh}(j)$; (g) $\operatorname{arccosh}(k)$; (h) $\operatorname{arctanh}(i)$;
   (i) $\operatorname{arctanh}(j)$; (j) $\operatorname{arctanh}(k)$.
2. Compute (a) $\operatorname{arctanh}(1+i)$; (b) $\operatorname{arctanh}(1+j)$; (c) $\operatorname{arctanh}(1+k)$;
   (d) $\operatorname{arctanh}(1+i+j+k)$; (e) $\operatorname{arctanh}(1-i-j+k)$; (f) $\operatorname{arctanh}(1-i-j-k)$.
3. Compute (a) $\ln(i) + \ln(1-i-j) + \sinh(i) + \operatorname{arctanh}(1-i-j-k)$;
   (b) $\ln(i)\ln(j) + (\sinh(i), \sinh(j)) + 2\operatorname{arctanh}(1+i)$; (c) $\ln(i) - \ln(k) + \sinh(i) \cdot \sinh(j) + \operatorname{arctanh}(1+j)$.
4. Compute (a) $\arcsin(i)$; (b) $\arcsin(j)$; (c) $\arcsin(k)$; (d) $\arccos(i)$;
   (e) $\arccos(j)$; (f) $\arccos(k)$.
5. Prove the following identities:

   (a) $\tanh(2\operatorname{arcsinh}(p)) = \left(2p\sqrt{p^2+1}\right)\left(2p^2+1\right)^{-1}$;

   (b) $\coth(2\operatorname{arcsinh}(p)) = \left(2p^2+1\right)\left(2p\sqrt{p^2+1}\right)^{-1}$;

   (c) $\tanh(2\operatorname{arccosh}(p)) = 2p\sqrt{p^2-1}\left(2p^2-1\right)^{-1}$;

   (d) $\coth(2\operatorname{arccosh}(p)) = \left(2p^2-1\right)\left(2p\sqrt{p^2-1}\right)^{-1}$;

   (e) $\sinh(3p) = \sinh(p)\left(4\cosh^2(p)-1\right)$;

   (f) $\sinh(3p) = \sinh(p)\left(4\sinh^2(p)+3\right)$;

   (g) $\sinh(3\operatorname{arccosh}(p)) = \sqrt{p^2-1}\left(4p^2-1\right)$;

   (h) $\cosh(3p) = \cosh(p)\left(4\cosh^2(p)-3\right)$;

   (i) $\cosh(3p) = \cosh(p)\left(4\sinh^2(p)+1\right)$;

   (j) $\cosh(3\operatorname{arcsinh}(p)) = \sqrt{p^2+1}\left(4p^2+1\right)$;

   (k) $\cosh(4p) = 8\cosh^2(p) - 8\cosh(p) + 1$;

   (l) $\sinh(4p) = 8\sinh(p)\cosh^3(p) - 4\sinh(p)\cosh(p)$;

   (m) $\sinh(4\operatorname{arccosh}(p)) = 4\sqrt{p^2-1}\,p\left(2p^2-1\right)$;

   (n) $\sinh(4\operatorname{arcsinh}(p)) = 4p\sqrt{p^2+1}\left(2p^2+1\right)$;

   (o) $\cosh(4\operatorname{arccosh}(p)) = 8p^2 - 8p + 1$;

   (p) $\cosh(4\operatorname{arcsinh}(p)) = 8\left(p^2+1\right) - 8\sqrt{p^2+1} + 1$;

   (q) $\tanh(4\operatorname{arcsinh}(p)) = 4p\sqrt{p^2+1}\left(2p^2+1\right)\left(8(p^2+1) - 8\sqrt{p^2+1}+1\right)^{-1}$;

   (r) $\coth(4\operatorname{arcsinh}(p)) = (8(p^2+1)-8\sqrt{p^2+1}+1)\left(4p\sqrt{p^2+1}(2p^2+1)\right)^{-1}$;

   (s) $\tanh(4\operatorname{arccosh}(p)) = 4\sqrt{p^2-1}\,p\left(2p^2-1\right)\left(8p^2-8p+1\right)^{-1}$;

   (t) $\coth(4\operatorname{arccosh}(p)) = (8p^2-8p+1)\left(4\sqrt{p^2-1}\,p(2p^2-1)\right)^{-1}$.

6. Let p be a real quaternion such that $|\mathbf{p}| = 0$, or

$$1 + 2p_0\sqrt{4p_0^2|\mathbf{p}|^2 - 3\left(p_0^2 + 1 - |\mathbf{p}|^2\right)^2} = 0.$$

   Prove that $\operatorname{arcsinh}(p) = \log_e|p_0 + C|$, where $C = p_0^2 + 1 - |\mathbf{p}|^2$.

7. Let p be a real quaternion such that $|\mathbf{p}| \neq 0$, or

$$1 + 2p_0 \sqrt{4p_0^2|\mathbf{p}|^2 - 3\left(p_0^2 - 1 - |\mathbf{p}|^2\right)^2} \neq 0.$$

Prove that

$$\mathrm{arccosh}(p) = \log_e |p_0 + C|$$
$$+ \mathrm{sgn}(\mathbf{p} + \mathrm{sgn}(B)\sqrt{A^2 - C^2})\left(\mathrm{arccos}\ \frac{p_0 + C}{|p_0 + C + \mathbf{p}\,\mathrm{sgn}(B)\sqrt{A^2 - C^2}|} + 2\pi n\right),$$

for $n \in \mathbb{Z}$, where

$$A := \sqrt{(p_0^2 - 1 - |\mathbf{p}|^2)^2 + 4p_0^2|\mathbf{p}|^2},$$

$$B := 2p_0\mathbf{p}, \quad C := p_0^2 - 1 - |\mathbf{p}|^2.$$

8. Let p be a real quaternion such that $|\mathbf{p}| = 0$, or

$$1 + 2p_0 \sqrt{4p_0^2|\mathbf{p}|^2 - 3\left(p_0^2 - 1 - |\mathbf{p}|^2\right)^2} = 0.$$

Prove that $\mathrm{arccosh}(p) = \log_e |p_0 + C|$, where $C = p_0^2 - 1 - |\mathbf{p}|^2$.

Quaternion Matrices

In a brief outline, the next portion of text describes a way of representing quaternion matrices in such a way that quaternionic addition and multiplication correspond to matrix addition, (scalar) matrix multiplication, and matrix transposition. Besides the discussion of the quaternionic analogues to complex matrices, we will also discuss the possibility of decomposing quaternions with respect to well known matrix decompositions, which are related to solving systems of linear equations. Even though it is self-contained, the reader will comprehend the necessity of this notions while pursuing the subject. Our presentation is organized in such a way that the analogies between quaternion, noncommutativity, and quaternion matrix, on one hand, and determinant, rank, and eigenvalues, on the other, are mastered. The relations between these concepts will come into view more clearly through the text, motivating our insight to the problem.

To treat this topic, we assume that you either have had some experience with matrices or are willing to learn something about them. Nowadays the study of quaternion matrices has undoubtedly gained an increased attention. For example, the representation of quaternions via matrices allows one to derive closed-form solutions for algebraic systems of linear equations involving unknown parameters, in which fast outputs and high level of data accuracy are of great concern. The set of equations for the quaternion elements is thus represented by a single equation. The solution can be expressed in terms of the same symbols, and matrix algebra knowledge gives us a routine procedure for finding the latter. Since matrices can be added (or subtracted), and multiplied together according to elementary algebraic techniques, they form a key tool to perform linear transformations.

9.1 Quaternion Matrices

Consider a set of quaternion-valued quantities arranged in a rectangular array containing m rows and n columns:

J.P. Morais et al., *Real Quaternionic Calculus Handbook*,
DOI 10.1007/978-3-0348-0622-0_9, © Springer Basel 2014

$$
\begin{pmatrix}
a_{1,1} & a_{1,2} & \dots & a_{1,n} \\
a_{2,1} & a_{2,2} & \dots & a_{2,n} \\
\vdots & \vdots & \ddots & \vdots \\
a_{m,1} & a_{m,2} & \dots & a_{m,n}
\end{pmatrix}.
$$

This array will be called an $m \times n$ quaternion matrix. The quantities $a_{i,j}$ ($i = 1,\dots,m$, $j = 1,\dots,n$) are called the entries or components of the matrix.

The collection of all $m \times n$ quaternion matrices with entries in $\mathbb{X}$ ($\mathbb{X} = \mathbb{R}$, $\mathbb{C}$ or $\mathbb{H}$) will be denoted by $\mathcal{M}_{m\times n}(\mathbb{X})$. For simplicity, in the case of square matrices ($m = n$) we denote $\mathcal{M}_{n\times n}(\mathbb{X})$ briefly by $\mathcal{M}_n(\mathbb{X})$. Throughout the remainder of the text we will point out remarkable similarities and differences between these type of matrices.

9.2   Equality

Two quaternion matrices A and B of same order $m \times n$ are said to be equal if all of their components are equal: $a_{i,j} = b_{i,j}$, for all $i = 1,\dots,m$, $j = 1,\dots,n$. We then write $A = B$. Two matrices of different order cannot be compared for equality.

9.3   Rank

The left (resp. right) rank of a quaternion matrix A is defined as the maximum number of columns of A that are left (resp. right) linearly independent, and is denoted by $r_l(A)$ (resp. $r_r(A)$). If A is of left (resp. right) rank k, then k is also the maximum number of rows of A that are left (resp. right) linearly independent.

Exercise 9.207. *Find an example of two quaternions that are left linearly independent, but not right linearly independent.*

Solution. $p = i, q = j$.

The relevant facts are that, just as with real and complex matrices, one can also define elementary row (and column) operations for quaternion matrices of compatible sizes.

9.4   Matrix Arithmetic and Operations

Quaternion matrices can be added, subtracted, and multiplied. If A and B are two quaternion matrices, and λ a real quaternion, these operations are defined as follows:

(i) The sum (resp. difference) $A + B$ (resp. $A - B$) of two matrices A and B of the same size is calculated by adding (resp. subtracting) entrywise;

(ii) The quaternion multiplication λA is given by multiplying every entry of A on the left-hand side by λ;

(iii) The multiplication of two quaternion matrices A and B, say AB, is defined only if the number of columns of A is the same as the number of rows of B. The matrix product AB is then given by the dot product of the corresponding row of A and the corresponding column of B.

The familiar associative, and distributive laws of multiplication over addition hold for quaternion matrices:

(iv) Commutative law of addition
$$A + B = B + A;$$

(v) Associative law of addition
$$A + (B + C) = (A + B) + C;$$

(vi) (Left-)distributive law of multiplication over addition
$$A(B + C) = AB + AC;$$

(vii) (Right-)distributive law of multiplication over addition
$$(B + C)A = BA + CA;$$

(viii) Associative law of multiplication
$$(AB)C = A(BC).$$

Exercise 9.208. *Let*

$$A = \begin{pmatrix} i & j \\ k & i \end{pmatrix} \quad \text{and} \quad A + B = \begin{pmatrix} 0_{\mathbb{H}} & 2j \\ 0_{\mathbb{H}} & 2i \end{pmatrix}.$$

Find B.

Solution. $B = \begin{pmatrix} -i & j \\ -k & i \end{pmatrix}.$

Exercise 9.209. *Let*

$$B = \begin{pmatrix} 1 & -i \\ j & -j \end{pmatrix}, \quad C = \begin{pmatrix} 1 & -i \\ i + k & j \end{pmatrix}, \quad A + B + C = \begin{pmatrix} 2 + 3i & -2i + j \\ i + 2j + 2k & i - 2j \end{pmatrix}.$$

Find A.

Solution. $A = \begin{pmatrix} 3i & j \\ j + k & i - 2j \end{pmatrix}.$

Exercise 9.210. *Let*

$$A = \begin{pmatrix} i & j \\ i & j \end{pmatrix}, \quad C = \begin{pmatrix} i & i \\ i & i \end{pmatrix}, \quad A(BC) = 2 \begin{pmatrix} -1 + k & -1 + k \\ -1 + k & -1 + k \end{pmatrix}.$$

Find B.

Solution. $B = \begin{pmatrix} k & k \\ k & k \end{pmatrix}.$

9.5   Special Quaternion Matrices

In this context, the additive identity and the multiplicative identity quaternion matrices will be denoted by O and I, respectively. The additive identity quaternion matrix, which acts as an annihilator element for quaternion matrix multiplication and as the additive identity for quaternion matrix addition, is such that for any quaternion matrix A, we have $A + O = A$, and $AO = OA = O$. Similarly, the multiplicative identity is such that for any quaternion matrix A, we have $AI = IA = I$.

As the next example shows, while the quaternion matrix addition is commutative, the quaternion matrix multiplication is not: for example, AB and BA are not equal in general. In case $AB = BA$, the quaternion matrices A and B are said to commute.

Example. Let

$$A = \begin{pmatrix} i & 0_\mathbb{H} \\ 0_\mathbb{H} & i \end{pmatrix} \quad \text{and} \quad B = \begin{pmatrix} j & 0_\mathbb{H} \\ 0_\mathbb{H} & j \end{pmatrix}.$$

Show that A and B do not commute.

Solution. One has that

$$AB = \begin{pmatrix} k & 0_\mathbb{H} \\ 0_\mathbb{H} & k \end{pmatrix} \neq \begin{pmatrix} -k & 0_\mathbb{H} \\ 0_\mathbb{H} & -k \end{pmatrix} = BA.$$

Exercise 9.211. *Let p be a real quaternion. Show that the following quaternion matrices commute:*

$$A_p = \begin{pmatrix} p & 1 \\ 0_\mathbb{H} & p \end{pmatrix} \quad \text{and} \quad B_p = \begin{pmatrix} \overline{p} & 0_\mathbb{H} \\ 0_\mathbb{H} & \overline{p} \end{pmatrix}$$

Exercise 9.212. *Let A be an $n \times n$ quaternion matrix for which it holds $A^2 + A + I = O$. Prove that*
(a) $A^3 = I$;
(b) $A^{2n} + A^n + I = \begin{cases} O, & \text{when } n \neq 3l \text{ for all } l \in \mathbb{N}, \\ 3I, & \text{when } n = 3l \text{ for some } l \in \mathbb{N}. \end{cases}$

9.6   **Basic Notions of Quaternion Matrices**

A quaternion square matrix A is said to be: idempotent if $A^2 = A$, Hermitian (or self-adjoint) if $\overline{A}^T = A$, skew-Hermitian (or anti-Hermitian) if $\overline{A}^T = -A$, unitary if $\overline{A}^T = A^{-1}$, normal if $A\overline{A}^T = \overline{A}^T A$, and invertible if $AB = BA = I$, for some matrix B. Two quaternion matrices A and B are said to be congruent if there exists an invertible matrix C such that $B = \overline{C}^T AC$. The conditions required in these definitions imply that we are dealing with square matrices. For a more unified formulation we emphasize that the symbol $\overline{A}$ means the quaternion conjugate of the matrix A, which is the matrix obtained by taking the conjugate of each entry in A. Clearly, $\overline{A}^T$ is the transpose of $\overline{A}$, that is the matrix obtained by interchanging the rows with the columns. In this sense, a quaternion hermitian matrix can be understood as the quaternionic extension of complex symmetric matrices. Just as for complex matrices, henceforth we denote $\overline{A}^T$ simply by A^*. The negative $-A$ is the matrix formed by changing the signs of all the entries of A, while the matrix A^{-1} is the multiplicative inverse of A. Quaternion skew-Hermitian matrices can be then understood as the quaternionic versions of complex skew-symmetric matrices.

Exercise 9.213. *Let p and q be two real quaternions, and*

$$A_{p,q} = \begin{pmatrix} 1 & p+q \\ 0_{\mathbb{H}} & 1 \end{pmatrix}.$$

When is the quaternion matrix $A_{p,q}$ idempotent? Is $A_{p,q}$ invertible?

Solution. $A_{p,q}$ is idempotent if $p + q = 0_{\mathbb{H}}$. $A_{p,q}$ is invertible.

Exercise 9.214. *Let $A = \begin{pmatrix} a & i \\ j & k \end{pmatrix}$, where $a \in \mathbb{H}$. Find a so that $A^2 = A$.*

Solution. No solutions.

Exercise 9.215. *Let $A = \begin{pmatrix} 0_{\mathbb{H}} & i \\ -i & 0_{\mathbb{H}} \end{pmatrix}$. Prove that A is Hermitian.*

Exercise 9.216. *Let p be a pure quaternion and A_p be the $n \times n$ quaternion matrix*

$$A_p = \begin{pmatrix} p & \cdots & p \\ \vdots & \ddots & \vdots \\ p & \cdots & p \end{pmatrix}.$$

Prove that (a) $A_p^{2l} = (-n|p|^2)^l I$ *for all $l \in \mathbb{N}$;*   (b) $A_p^{2l+1} = (-n|p|^2)^l A_p$ *for all $l \in \mathbb{N}$.*

The inverse of quaternion matrices deserves particular interest, and so we will now study this in more detail.

9.7   Inverse of a Quaternion Matrix

Let $A = A_1 + A_2 j$ be a quaternion matrix with an inverse $A^{-1} = T_1 + T_2 j$, where A_1, A_2, T_1, and T_2 are all complex. Next we show how to find A^{-1} from A_1 and A_2 by inverting complex matrices. The detailed process can be described as follows:
1. Write A as $A_1 + A_2 j$, where A_1 and A_2 are complex matrices;
2. Form the complex matrix

$$A_c = \begin{pmatrix} A_1 & A_2 \\ -\overline{A_2} & \overline{A_1} \end{pmatrix};$$

3. Find A_c^{-1} in the classical sense,

$$A_c^{-1} = \begin{pmatrix} T_1 & T_2 \\ -\overline{T_2} & \overline{T_1} \end{pmatrix};$$

4. $A^{-1} := T_1 + T_2 j$.

Example. Compute the inverse of the quaternion matrix

$$A = \begin{pmatrix} i + j & i - j \\ 0_{\mathbb{H}} & i - j \end{pmatrix}.$$

Solution. First, write A as

$$A = \begin{pmatrix} i & i \\ 0_{\mathbb{H}} & i \end{pmatrix} + \begin{pmatrix} j & -j \\ 0_{\mathbb{H}} & -j \end{pmatrix}$$

$$= \begin{pmatrix} i & i \\ 0_{\mathbb{H}} & i \end{pmatrix} + \begin{pmatrix} 1_{\mathbb{H}} & -1_{\mathbb{H}} \\ 0_{\mathbb{H}} & -1_{\mathbb{H}} \end{pmatrix} j.$$

It follows that

$$A_1 = \begin{pmatrix} i & i \\ 0_{\mathbb{H}} & i \end{pmatrix}, \quad \text{and} \quad A_2 = \begin{pmatrix} 1_{\mathbb{H}} & -1_{\mathbb{H}} \\ 0_{\mathbb{H}} & -1_{\mathbb{H}} \end{pmatrix}$$

and

$$A_c = \begin{pmatrix} i & i & 1_{\mathbb{H}} & -1_{\mathbb{H}} \\ 0_{\mathbb{H}} & i & 0_{\mathbb{H}} & -1_{\mathbb{H}} \\ -1_{\mathbb{H}} & 1_{\mathbb{H}} & -i & -i \\ 0_{\mathbb{H}} & 1_{\mathbb{H}} & 0_{\mathbb{H}} & -i \end{pmatrix}.$$

Then it is easily seen that $\det(A_c^{-1}) = 4$, and after some straightforward computations we find

$$A_c^{-1} = \frac{1}{4} \begin{pmatrix} -2i & 2i & -2 & 2 \\ 0_{\mathbb{H}} & -2i & 0_{\mathbb{H}} & 2 \\ 2 & -2 & 2i & -2i \\ 0_{\mathbb{H}} & -2 & 0_{\mathbb{H}} & 2i \end{pmatrix}.$$

We conclude that

$$T_1 = \frac{1}{4} \begin{pmatrix} -2i & 2i \\ 0_{\mathbb{H}} & -2i \end{pmatrix} \quad \text{and} \quad T_2 = \frac{1}{4} \begin{pmatrix} -2 & 2 \\ 0_{\mathbb{H}} & 2 \end{pmatrix}.$$

Hence the inverse matrix is given by

$$A^{-1} = T_1 + T_2 j$$

$$= \frac{1}{4} \begin{pmatrix} -2i - 2j & 2i + 2j \\ 0_{\mathbb{H}} & -2i + 2j \end{pmatrix}.$$

We continue by pointing out some properties of the conjugate and the transpose of a quaternion matrix, some of which are unexpected.

9.8　Quaternion Conjugate of a Matrix

Let A and B be two quaternion matrices and λ a real quaternion. The quaternion conjugate has the following properties:

(i) $\overline{A + B} = \overline{A} + \overline{B}, \quad \overline{\lambda A} = \overline{A}\,\overline{\lambda};$　　(ii) $\overline{AB} \neq \overline{B}\,\overline{A}$ in general;

(iii) $\overline{\overline{A}} = A;$　　(iv) $\overline{A^{-1}} \neq (\overline{A})^{-1}$ in general.

Property (i) can be extended to three or more quaternion matrices, i.e., $\overline{A_1 + A_2 + \cdots + A_n} = \overline{A_1} + \overline{A_2} + \cdots + \overline{A_n}$, i.e., the conjugate of a sum of quaternion matrices is the sum of the conjugates of the terms in the sum. However, the equalities $\overline{AB} = \overline{B}\,\overline{A}$ and $\overline{A^{-1}} = (\overline{A})^{-1}$ do not hold in general. To see, say, that (iv) holds, take for example

$$A = \begin{pmatrix} i & k \\ 0_{\mathbb{H}} & j \end{pmatrix} \quad \text{and} \quad A^{-1} = \begin{pmatrix} -i & -1_{\mathbb{H}} \\ 0_{\mathbb{H}} & -j \end{pmatrix}.$$

Exercise 9.217. *Let A be a nonnull quaternion matrix and λ a real quaternion. Under what circumstances does the equality $\overline{\lambda A} = \lambda \overline{A}$ hold?*

Exercise 9.218. *Let $A = \begin{pmatrix} i & j \\ k & 1_{\mathbb{H}} \end{pmatrix}$ and $B = \begin{pmatrix} i & i \\ i & i \end{pmatrix}$. Check if $\overline{AB} = \overline{B}\,\overline{A}$.*

9.9 Quaternion Transpose, Conjugate Transpose, and Inverse Matrices

Let A and B be two quaternion matrices and λ a real quaternion. Here is a list of immediate properties:

(i) Linearity
$$(A + B)^T = A^T + B^T, \quad (\lambda A)^T = \lambda A^T;$$

(ii) Product
$$(AB)^T \neq B^T A^T \text{ in general}, \quad (AB)^* = B^* A^*;$$

(iii) Involutivity
$$(A^T)^T = A, \quad (\overline{A})^T = \overline{(A^T)};$$

(iv) Inverse
$$(A^{-1})^T \neq (A^T)^{-1} \text{ in general},$$
$$(AB)^{-1} = B^{-1} A^{-1} \text{ if } A \text{ and } B \text{ are invertible},$$
$$(A^*)^{-1} = (A^{-1})^* \text{ if } A \text{ is invertible};$$

(v) For each A there exists a unitary matrix U such that $U^* A U$ is in upper triangular form [7].

Property (i) can be extended to the general case of an arbitrary number of quaternion matrices. For example, $(A_1 + A_2 + \ldots + A_n)^T = A_1^T + A_2^T + \ldots + A_n^T$. On the contrary, $(A_1 A_2 \cdots A_{n-1} A_n)^T = A_n^T A_{n-1}^T \cdots A_2^T A_1^T$ does in general not hold for quaternion matrices. In words, the transpose of a sum of quaternion matrices is the sum of the transposes of the terms in the sum. The transpose of a product of quaternion matrices is in general not equal to the product of the transposes of each factor in the reverse order.

Exercise 9.219. *Check if* $(AB)^T = B^T A^T$, *where*

$$A = \begin{pmatrix} i & j & k \\ -i & i+j & -k \\ 1-i-j & k & k \end{pmatrix} \quad \text{and} \quad B = \begin{pmatrix} 1+i & j & -k \\ 2j & -k & -i \\ i & k & i \end{pmatrix}.$$

Exercise 9.220. *Let A and B be two quaternion matrices. Check if* $(\overline{AB})^T = (\overline{B})^T (\overline{A})^T$.

Exercise 9.221. *Let A be a quaternion matrix. Find a counterexample to show that the equality* $(A^{-1})^T = (A^T)^{-1}$ *does not hold in general.*

Exercise 9.222. *Prove that any Hermitian quaternion matrix is congruent to a real diagonal matrix.*

9.10 Positive and Semi-positive Definite Quaternion Matrices

If A is a Hermitian $n \times n$ quaternion matrix, then X^*AX is a real number for any n-dimensional column vector X with quaternion entries. A Hermitian quaternion matrix A is called positive definite (resp. semi-positive definite) if X^*AX is positive (resp. nonnegative) for any nonzero X. We have a list of properties:

  (i) If A is positive definite (resp. semi-positive definite), then the elements on the diagonal of A are all positives (resp. nonnegative);
 (ii) If A is diagonal, then A is positive definite (resp. semi-positive definite) if and only if the elements on the diagonal of A are all positive (resp. nonnegative);
(iii) If Hermitian matrices A and B are congruent, then A is positive definite (resp. semi-positive definite) if and only if B is positive definite (resp. semi-positive definite);
(iv) A Hermitian quaternion matrix A is semi-positive definite if and only if A is congruent to a matrix of the form $\begin{pmatrix} I & O \\ O & O \end{pmatrix}$.

Exercise 9.223. *Let A, B be two Hermitian $n \times n$ quaternion matrices, and k be any fixed real number. Prove that A, B can be simultaneously diagonalized by congruence if and only if $A, kA + B$ can be simultaneously diagonalized by congruence.*

Exercise 9.224. *Let A, B be two Hermitian $n \times n$ quaternion matrices. Prove that if one of them is positive definite, then there exists an invertible matrix C such that both C^*AC and C^*BC are diagonal.*

9.11 Determinant of a Quaternion Matrix

Associated with any square quaternion matrix is a real number, which we now define. The determinant of a quaternion square matrix A is a quaternion number, denoted by $\det(A)$. It is significant to note that, in practice, the determinant of a 2×2 matrix is defined by

$$\det \begin{pmatrix} a_{11} & a_{12} \\ a_{21} & a_{22} \end{pmatrix} = a_{11}a_{22} - a_{12}a_{21}.$$

In the above definition we use the so-called rule "multiplication from above to down below". In the same way, we may use the rule "multiplication from down below to above" i.e.

$$\det \begin{pmatrix} a_{11} & a_{12} \\ a_{21} & a_{22} \end{pmatrix} = a_{22}a_{11} - a_{21}a_{22}.$$

For the reader's convenience we will use the rule "multiplication from above to down below" only. The other case can be treated analogously. To proceed with, the determinant of a 3×3 matrix is defined by

$$\det \begin{pmatrix} a_{11} & a_{12} & a_{13} \\ a_{21} & a_{22} & a_{23} \\ a_{31} & a_{32} & a_{33} \end{pmatrix} = a_{11} \det \begin{pmatrix} a_{22} & a_{23} \\ a_{32} & a_{33} \end{pmatrix} - a_{12} \det \begin{pmatrix} a_{21} & a_{23} \\ a_{31} & a_{33} \end{pmatrix}$$

$$+ a_{13} \det \begin{pmatrix} a_{21} & a_{22} \\ a_{31} & a_{32} \end{pmatrix}$$

$$= a_{11}a_{22}a_{33} + a_{12}a_{23}a_{31} + a_{13}a_{21}a_{32}$$

$$- a_{11}a_{23}a_{32} - a_{12}a_{21}a_{33} - a_{13}a_{22}a_{31}.$$

Example. Compute the determinant of the 3×3 quaternion matrix

$$\begin{pmatrix} i & j & 0_{\mathbb{H}} \\ 0_{\mathbb{H}} & -j & k \\ i & j & k \end{pmatrix}.$$

Solution. A straightforward computation shows that

$$\det \begin{pmatrix} i & j & 0_{\mathbb{H}} \\ 0_{\mathbb{H}} & -j & k \\ i & j & k \end{pmatrix} = i \det \begin{pmatrix} -j & k \\ j & k \end{pmatrix} - j \det \begin{pmatrix} 0_{\mathbb{H}} & k \\ i & k \end{pmatrix}$$

$$= i(-jk - kj) - j(0_{\mathbb{H}} - ki)$$

$$= -1_{\mathbb{H}}.$$

Exercise 9.225. *Let* $A = \begin{pmatrix} a & i+k \\ i-j & i-k \end{pmatrix}$, *where* $a \in \mathbb{H}$. *Find* a *so that* $\det(A - iI) = 0_{\mathbb{H}}$.

Solution. $a = 1 + 2i - j - k$.

Exercise 9.226. *Let* $A = \begin{pmatrix} a & i \\ i+j & k \end{pmatrix}$, *where* $a \in \mathbb{H}$. *Find* a *so that* $\det(A) = 1 - k$.

Solution. $a = 0_{\mathbb{H}}$.

Exercise 9.227. *Let* $A = \begin{pmatrix} a & b \\ i+j & k \end{pmatrix}$, *where* $a, b \in \mathbb{H}$. *Find* a *and* b *so that* $\det(A) = i$.

Solution. $a = (b+1)j$.

In general, we compute the determinant of an $n \times n$ quaternion matrix in the following way:

$$\det \begin{pmatrix} a_{11} & a_{12} & \cdots & a_{1n} \\ a_{21} & a_{22} & \cdots & a_{2n} \\ \vdots & \vdots & \ddots & \vdots \\ a_{n1} & a_{n2} & \cdots & a_{nn} \end{pmatrix}$$

$$= a_{11} \det \begin{pmatrix} a_{22} & \cdots & a_{2n} \\ \vdots & \ddots & \vdots \\ a_{n2} & \cdots & a_{nn} \end{pmatrix} - \cdots + (-1)^{1+n} a_{1n} \det \begin{pmatrix} a_{21} & \cdots & a_{2n-1} \\ \vdots & \ddots & \vdots \\ a_{n1} & \cdots & a_{nn-1} \end{pmatrix}.$$

Let A and B be two quaternion matrices. The following properties hold:
(i) A is invertible if and only if $\det(A) \neq 0_{\mathbb{H}}$;
(ii) $\det(AB) = \det(A)\det(B)$, consequently $\det(A^{-1}) = \det(A)^{-1}$, if A^{-1} exists;
(iii) $\det(PAQ) = \det(A)$, for any elementary quaternion matrices P and Q.

Exercise 9.228. *Let A and B be two quaternion matrices. Find an example for which* $\det(A^3 + B) = \det(A) + \det(B)$.

Solution. Let p be a pure unit quaternion. Take $A = -\begin{pmatrix} p & p \\ p & p \end{pmatrix}$ and $B = O$.

9.12 Dieudonné Determinant

Let $A = \begin{pmatrix} a & b \\ c & d \end{pmatrix}$ be a 2×2 matrix with quaternion entries. The Dieudonné determinant of A is defined to be the nonnegative real number

$$\det_{\mathbb{H}}(A) := \sqrt{|a|^2|d|^2 + |c|^2|b|^2 - 2\mathrm{Sc}(c\overline{a}b\overline{d})}.$$

Exercise 9.229. *Compute the Dieudonné determinant of the 2×2 quaternion matrix* $\begin{pmatrix} i-j & j-2k \\ -j+k & i+k \end{pmatrix}$.

Solution. $\sqrt{14}$.

Exercise 9.230. *Let* $A = \begin{pmatrix} a & b \\ c & d \end{pmatrix}$ *be a* 2×2 *matrix with quaternion entries. Prove that* $\det_{\mathbb{H}}(A) \leq |a||d| + |c||b|$.

Exercise 9.231. *Let* $A = \begin{pmatrix} a & b \\ c & d \end{pmatrix}$ *be a* 2×2 *matrix with quaternion entries. Prove that*
(a) *If* $(a\overline{c}) \cdot (b\overline{d}) = 0$, *then* $\det_{\mathbb{H}}(A) = \sqrt{|a|^2|d|^2 + |c|^2|b|^2}$;
(b) *If* $(c\overline{a}) \cdot (d\overline{b}) = 0$, *then* $\det_{\mathbb{H}}(A) = \sqrt{|a|^2|d|^2 + |c|^2|b|^2}$.

9.13   Quaternion Eigenvalues and Eigenvectors

For quaternion matrices an eigenvalue theory can be developed similar to that for complex matrices. However, due to the noncommutativity of multiplication, we need to treat $Ax = x\lambda$ and $Ax = \lambda x$ separately. In the quaternionic case the following generalization holds. Let A be a quaternion matrix and consider the equation

$$Ax = x\lambda_{\mathrm{r}} \quad (\text{resp. } Ax = \lambda_{\mathrm{l}}x), \tag{9.1}$$

where λ_{r} (resp. λ_{l}) is a quaternion and x is a quaternion vector. If (9.1) has a nonzero solution x, then λ_{r} (resp. λ_{l}) is called a right (resp. left) characteristic root or eigenvalue of the matrix A and x is called a corresponding quaternion eigenvector.

Moreover, we state the following results:
 (i) Every $n \times n$ quaternion matrix has at least one left eigenvalue in $\mathbb{H}$ [60],
 (ii) Any $n \times n$ quaternion matrix has exactly n right eigenvalues, which are complex numbers with nonnegative imaginary parts [7, 33],
 (iii) If $A \in \mathcal{M}_{m \times n}(\mathbb{H})$, with $m < n$, then $Ax = 0_{\mathbb{H}}$ has a nonzero solution,
 (iv) If $A \in \mathcal{M}_n(\mathbb{H})$ is in triangular form, then every diagonal element is a right eigenvalue of A. Conversely, every right eigenvalue of A is similar to a diagonal entry of A [7].

Exercise 9.232. *Let* $A = \begin{pmatrix} i & b \\ j+k & k \end{pmatrix}$, *where* $b \in \mathbb{H}$. *Find* b *so that*

$$A \begin{pmatrix} x_1 \\ x_2 \end{pmatrix} = j \begin{pmatrix} x_1 \\ x_2 \end{pmatrix},$$

where $x_1, x_2 \in \mathbb{H}$ *are arbitrary.*

Solution. $b = \frac{1}{2} - \frac{1}{2}k.$

9.14 Spectrum of a Quaternion Matrix

The set $\{\lambda_r \in \mathbb{H} : Ax = x\lambda_r,$ for some $x \neq 0\}$ is called the right spectrum of A, denoted by $\sigma_r(A)$. The left spectrum is defined similarly and is denoted by $\sigma_l(A)$.

The following example illustrates how we calculate the (left) eigenvalues of a given quaternion matrix.

Example. Find the left spectrum of the matrix $A = \begin{pmatrix} 0_{\mathbb{H}} & -1_{\mathbb{H}} \\ 1_{\mathbb{H}} & 0_{\mathbb{H}} \end{pmatrix}$.

Solution. First we write $Ax = \lambda_l x$ as $(\lambda_l I - A)x = 0$, and assume that $\lambda_l I - A$ is invertible for all $\lambda_l \in \mathbb{H}$. The left spectrum of A is then the set of quaternionic solutions of $\lambda_l^2 + 1_{\mathbb{H}} = 0_{\mathbb{H}}$, which is a two-dimensional sphere.

Exercise 9.233. *Find the left eingenvalues of the following matrices:*

$$\begin{pmatrix} i & 0_{\mathbb{H}} \\ 0_{\mathbb{H}} & i \end{pmatrix}, \quad \begin{pmatrix} j & 0_{\mathbb{H}} \\ 0_{\mathbb{H}} & j \end{pmatrix}, \quad \begin{pmatrix} 0_{\mathbb{H}} & i \\ -i & 0_{\mathbb{H}} \end{pmatrix}, \quad \begin{pmatrix} 0_{\mathbb{H}} & i \\ j & 0_{\mathbb{H}} \end{pmatrix}.$$

Exercise 9.234. *Let A be a matrix with real entries and q be a unit quaternion. Prove that if λ_l is a left eigenvalue of A, then $\overline{q}\lambda_l q$ is also a left eigenvalue of A. What can you conclude if A has quaternion-valued entries?*

Exercise 9.235. *Let A be a quaternion matrix and let $\lambda_r \in \mathbb{H}$ be a right eigenvalue of A. If $\rho \in \mathbb{H} \setminus \{0_{\mathbb{H}}\}$, then $\rho^{-1}\lambda_r \rho$ is also a right eigenvalue of A.*

Exercise 9.236. *Let A be an upper triangular quaternion matrix. Prove that a quaternion λ_l is a left eigenvalue of A if and only if λ_l is a diagonal entry.*

To continue the exposition, we present the following results:

9.15 Right Eigenvalue Equalities

Let A be a quaternion matrix, λ_r one of its right eigenvalues, and x a corresponding vector. Let μ_r be a right eigenvalue of $\overline{A}^T$, with y a corresponding vector. Then

$$\overline{\lambda_r}\, \overline{x}^T y = \overline{x}^T y\, \mu_r.$$

Let us prove this relation. By assumption $Ax = x\lambda_r$, and $\overline{A}^T y = y\mu_r$, and since the equality $(\overline{AB})^T = (\overline{B})^T(\overline{A})^T$ holds for two quaternion matrices A and B, we have

$$\overline{\lambda_r}\,\overline{x}^T y = (\overline{x}^T\overline{A}^T)y = \overline{x}^T(\overline{A}^T y) = \overline{x}^T y\, \mu_r.$$

Using the above relation, it can be proved that a quaternion Hermitian matrix has real right eigenvalues only. For the proof, let λ_r be a right eigenvalue of A with x as a corresponding vector. It is clear that λ_r is also a right eigenvalue of $\overline{A}^T$ $(= A)$ with vector x. From the relation above it follows that

$$\overline{\lambda_r}\, \overline{x}^T x \; = \; \overline{x}^T x \, \lambda_r.$$

Since $\overline{x}^T x$ is a real number different from zero, we have $\lambda_r = \overline{\lambda_r}$.

Exercise 9.237. *Two quaternion square matrices A and B are said to be similar if there exists an invertible quaternion matrix S of the same size such that $S^{-1}AS = B$. Prove that then A and B have the same right eigenvalues.*

9.16 Advanced Practical Exercises

1. Compute

(a) $j \begin{pmatrix} 1_{\mathbb{H}} & i \\ 1 + \frac{1}{2}i + j & k \end{pmatrix}$;   (b) $\begin{pmatrix} 1_{\mathbb{H}} & i \\ 1 + \frac{1}{2}i + j & k \end{pmatrix} j$;

(c) $k \begin{pmatrix} 1 - i + j + k & i & j\ k \\ i & 1 + i - j - k & j\ k \\ 0_{\mathbb{H}} & 2 - i + j + k\ i & j \end{pmatrix}$;

(d) $\begin{pmatrix} 1 - i + j + k & i & j\ k \\ i & 1 + i - j - k & j\ k \\ 0_{\mathbb{H}} & 2 - i + j + k\ i & j \end{pmatrix} k$.

2. Compute

$$\overline{1+i} \begin{pmatrix} 1_{\mathbb{H}} & i \\ -j & 1 + i + j + k \end{pmatrix} - 2j \begin{pmatrix} 2 & -3i \\ i & 4 + j \end{pmatrix}.$$

3. Find AB, where

(a) $A = \begin{pmatrix} 0_{\mathbb{H}} & -i & j \\ 1 + i & j & k \\ -i & 2 + i + j & i \end{pmatrix}$, $B = \begin{pmatrix} 2 & -i & j \\ i + k & j & i \\ i + j & k & 5 \end{pmatrix}$;

(b) $A = \begin{pmatrix} 2 + i & j \\ 1 - i - j + k & i + j \end{pmatrix}$, $B = \begin{pmatrix} i \\ j \end{pmatrix}$.

4. Find W^{-1}, where

$$\text{(a)}\quad W = \begin{pmatrix} j & -3k \\ -k & 1+j+k \end{pmatrix};\qquad \text{(b)}\quad W = \begin{pmatrix} i & j \\ 1_{\mathbb{H}} & -k \end{pmatrix};\qquad \text{(c)}\quad W = \begin{pmatrix} k & i \\ j & j \end{pmatrix}.$$

Check if $WW^{-1} = W^{-1}W = I$.

5. Find a vector X with quaternionic entries which solves the system $AX = B$, where

$$\text{(a)}\quad A = \begin{pmatrix} 1-i+2j & k \\ -i & j \end{pmatrix},\qquad B = \begin{pmatrix} k \\ j \end{pmatrix};$$

$$\text{(b)}\quad A = \begin{pmatrix} 1+i & j & k \\ i & 1+j & 1-k \\ j & i & k \end{pmatrix},\qquad B = \begin{pmatrix} 1_{\mathbb{H}} \\ i \\ j \end{pmatrix}.$$

6. Prove that every matrix with quaternionic entries has eigenvalues.

7. Prove that similar matrices have the same characteristic roots.

8. Find $x \in \mathbb{H}$ such that (a) $(1+i)x = xj$; (b) $(1-k)x = xi$;
(c) $(1+i+j+k)x = x(1-i+j+k)$.

9. Let $a, b \in \mathbb{H}\setminus\{0_{\mathbb{H}}\}$. Prove that the linear equation $ax = xb$ has a nonzero solution $x \in \mathbb{H}$ if and only if $a_0 = b_0$ and $|\mathbf{a}| = |\mathbf{b}|$.

10. Check that the following equations have nonzero solutions: (a) $(1+i)x = xj$;
(b) $(1-k)x = xi$; (c) $(1+i+j+k)x = x(1-i+j+k)$;
(d) $(1+2i+j+k)x = x(1+i+j+k)$; (e) $(2i+j)x = x(i+j)$;
(f) $ix = x(i+j)$.

11. Let $h_1, h_2 \in \mathbb{H}\setminus\{0_{\mathbb{H}}\}$ be chosen such that $h_1 a = a h_1$ and $h_2 b = b h_2$. Further, let y be a nonzero solution to the equation $ax = xb$. Prove that $h_1 y h_2$ is also a nonzero solution of $ax = xb$.

Monomials, Polynomials and Binomials 10

In this chapter we will be primarily interested in the study of monomials and polynomials within the framework of quaternion analysis. Monomials and their applications to combinatorics and number theory have become increasingly important for the study of a large number of problems that arise in many different contexts, both from a theoretical and a practical perspective. At the same time, the applications of polynomials to classical and numerical analysis, including approximation theory, statistics, combinatorics, number theory, group representations etc., as well as in physics, including quantum mechanics and statistical physics, and in system theory and signal processing have played a key role in this development and continue to do it today. For example, polynomials are often used in the treatment of problems, mainly in mathematical physics, and also in studies related to differential equations, continued fractions, and numerical stability. In advanced mathematics, polynomials are used to construct polynomial rings, a central concept in abstract algebra and algebraic geometry.

In the remainder of the chapter we are going to introduce into the realm of quaternion numbers the concept of binomial coefficient, which can be profitably thought as one of the most important combinatorial counting tools. Central to this viewpoint are certain binomial sums involving quaternions, which as in the real and complex cases, should be familiar to the reader. We restrict ourselves mainly to the most important special binomial coefficients in which the novelty lies in the arrangement, and also in some of the proofs.

We hope that after studying these examples the reader will be at least partly convinced of the power of the method, as well as of the beauty of the unified approach. Every effort has been made to make this chapter self-contained. In particular, we do think that a completely unified treatment of the topic is neither possible nor desirable.

J.P. Morais et al., *Real Quaternionic Calculus Handbook*,
DOI 10.1007/978-3-0348-0622-0_10, © Springer Basel 2014

10.1   Quaternion Monomials

For real variables t, x, y, and z, the term $p := t + xi + yj + zk$ is called a quaternion variable. Let n be a nonnegative integer. A quaternion monomial M_n of degree n is defined as $a_0 p\, a_1 p \cdots a_{n-1} p a_n$, where $a_l \in \mathbb{H}$ ($l = 0, 1, \ldots, n$) and such that $M_n(0) = 0_{\mathbb{H}}$. The degree (or order) of a quaternion monomial is the sum of the exponents of the powers of p. For example, $(1 + i) p (1 + j + k)\, pk$ is a quaternion monomial of degree 2. Two quaternion monomials M_1 and M_2 are said to be similar, if there exist two real quaternions a and b such that $M_1 = a M_2 b$ and the degree of M_1 is equal to the degree of M_2.

Example. Define $g(p) = p^n$ for $p \in \mathbb{H}$, and also set $g(\infty) = \infty$. Prove that the monomial g has exactly degree n.

Solution. We shall use induction on the degree of g in order to prove our statement. To start with, note that for $n = 1$ the assertion is evident, i.e. $p = b_0 q b_1$ where b_0, b_1 are real quaternions. Suppose that our assertion is valid for all n: $p^n = a_0 q a_1 \cdots q a_n$. We shall now prove that it is valid for $n + 1$. A direct computation shows that

$$p^{n+1} = p^n p = a_0 q a_1 \cdots q a_n p = a_0 q a_1 \cdots q a_n b_0 q b_1.$$

Denoting $a_n b_0 = c_1$, we conclude that $p^{n+1} = a_0 q a_1 \ldots q c_1 q b_1$. From the definition of the quaternion monomial it follows that p^{n+1} is a monomial of degree $n + 1$.

Exercise 10.238. *Determine the degree of the following quaternion monomials:* (a) $(1 + i + j)pj$;   (b) $(1 - i - j)pkpjp(1 + k)$;   (c) $(1 - 2i + 3k)p(1 + j)p(1 - i - j)p(1 + j + k)p(1 + i + j)$;   (d) $(1 - i - k)p(1 - i - j)p(1 + k)p(1 + i + 2k)p$.

Solution. (a) 1;   (b) 3;   (c) 4;   (d) 4.

Exercise 10.239. *Show that the following quaternion monomials are similar:* (a) pj and $(1 + i + j)pj$;   (b) $(1 - i)pjp$ and $(1 - i)pjpk$;   (c) $(1 - i - j - k)pjp(1 - j)p(1 + k)$ and $pjp(1 - j)p$.

Exercise 10.240. *Show that the following quaternion monomials are not similar:* (a) $(1 - i - j - k)pipjpkpjpi$ and ipk; (b) $(1 - i - k)pk$ and $ipipipjpkpjpipipikp$;   (c) $(i + j + k)pipkpjp$ and ip.

10.2   Quaternion Monomials Arithmetic and Operations

Quaternion monomials can be added, subtracted, multiplied, and divided. Let M_{n_1} and M_{n_2} be two quaternion monomials, with n_1 and n_2 nonnegative integers, and λ a real quaternion. These operations are defined as follows:

(i)   The sum (resp. difference) $M_{n_1} + M_{n_2}$ (resp. $M_{n_1} - M_{n_2}$) of two monomials M_{n_1} and M_{n_2} is calculated by adding (resp. subtracting) the coefficients of the similar terms;

(ii)   The quaternion multiplication λM_{n_1} is given by $\lambda(a_0 p a_1 p \cdots a_{n-1} p a_n) = (\lambda a_0) p a_1 \cdots a_{n-1} p a_n$;

(iii)   The multiplication of two quaternion monomials M_{n_1} and M_{n_2}, say $M_{n_1} M_{n_2}$, is the standard product of quaternions;

(iv)   The right- (resp. left-) quotient of two quaternion monomials M_{n_1} and M_{n_2}, say $M_{n_1} M_{n_2}^{-1}$ (resp., $M_{n_2}^{-1} M_{n_1}$), is the standard quotient of quaternions.

The familiar associativity, and distributivity laws of multiplication over addition hold for quaternion monomials:

(v)   Commutativity law of addition
$$M_{n_1} + M_{n_2} = M_{n_2} + M_{n_1} \quad \text{for all } M_{n_1}, M_{n_2} \in \mathbb{H};$$

(vi)   Associativity law of addition
$$M_{n_1} + (M_{n_2} + M_{n_3}) = (M_{n_1} + M_{n_2}) + M_{n_3} \quad \text{for all } M_{n_1}, M_{n_2}, M_{n_3} \in \mathbb{H};$$

(vii)   (Left-)distributivity law of multiplication over addition
$$M_{n_1}(M_{n_2} + M_{n_3}) = M_{n_1} M_{n_2} + M_{n_1} M_{n_3} \quad \text{for all } M_{n_1}, M_{n_2}, M_{n_3} \in \mathbb{H};$$

(viii)   (Right-)distributivity law of multiplication over addition
$$(M_{n_2} + M_{n_3}) M_{n_1} = M_{n_2} M_{n_1} + M_{n_3} M_{n_1} \quad \text{for all } M_{n_1}, M_{n_2}, M_{n_3} \in \mathbb{H};$$

(ix)   Associativity law of multiplication
$$(M_{n_1} M_{n_2}) M_{n_3} = M_{n_1}(M_{n_2} M_{n_3}) \quad \text{for all } M_{n_1}, M_{n_2}, M_{n_3} \in \mathbb{H}.$$

Exercise 10.241. *Let p be a quaternion variable. Find a solution p to the following equations:* (a) $p^2 - ipj = 1 - k;$ (b) $i(kpi - 2ip^3)j = 1_{\mathbb{H}}.$

Solution. No solutions.

Exercise 10.242. *Let p be a quaternion variable. Compute the following quaternion monomials:* (a) $(1 - i + j + k)pp(-i)i;$ (b) $\frac{1}{3}(i + jp^5);$ (c) $(-jpk)^9;$ (d) $\frac{1}{n!}p^{n-1}p(-k)k$ *for a nonnegative integer n.*

Solution. (a) $(1 - i + j + 3k)p^2;$ (b) $\frac{1}{3}i + \frac{1}{3}jp^5;$ (c) $-jpipipipipipipipip;$ (d) $\frac{1}{n!}p^n.$

Exercise 10.243. *Let $M_1 = ipj + (1 - i - j)pipj$, $M_2 = pj + pkp$ and $M_3 = kp.$ Find* (a) $M_1 + M_2;$ (b) $M_1 - M_2;$ (c) $M_1 M_3.$

Solution. (a) $(1+i)pj+(1-i-j)pipj+pkp$;   (b) $(-1+i)pj+(1-i-j)pipj+pkp$;
(c) $ipip + (1 - i - j)pipip$.

As a brief preview of the material presented in a further chapter, we proceed by
recalling the notion of quaternion polynomial. Until we reach this point the reader
is asked to subdue his quest for complete rigor.

10.3   Quaternion Polynomials

Let p be a quaternion variable, n and m nonnegative integers. A quaternion polynomial $P_n(p)$ of degree n is a finite linear combination of quaternion monomials:
$a_0\, pa_1 \cdots pa_n + M_m(p)$, where $a_l \in \mathbb{H}$ $(l = 0, 1, \ldots, n)$ and $M_m(p)$ is a finite sum
of quaternion monomials of degree $m \leq n$. The degree (or order) of a quaternion
polynomial is the largest of the degrees of the monomials, forming this polynomial.
For example, $(2 + k)pjp - pj + ip + k$ is a quaternion polynomial of degree 2.
Notice that neither $t - \frac{1}{x}i + yzk$ nor $7yz^{1/2}j$ are quaternion polynomials, since the
first one involves division by the variable x and the second one contains an exponent
that is not an integer. There are situations in which the sum (difference), the product
and division of quaternion polynomials are again quaternion polynomials.

Exercise 10.244. *Determine the degree of the following quaternion polynomials:*
(a) $g(p) = ipk + (1 + i)p(1 + j)p + kpipj$;   (b) $g(p) = i + (1 - i - j)p(1 -
k) + (1 - i - j - k)pkp(1 - i)p + ipk$.

Solution. (a) 2;   (b) 3.

Exercise 10.245. *Find a polynomial* $q(p) = (p - a)(p - i)$ *such that* $q(i) = 0$
and $q(j) = -1 - 2j + 3k$.

Solution. $q(p) = (p - 1 + i + j + k)(p - i)$.

Exercise 10.246. *Find a polynomial* $g(p) = (p - b)(ipk - j)pi$ *such that* $g(k) =
2j + 2k$.

Solution. $g(p) = (p - 1 + i + j)(ipk - j)pi$.

10.4   Factoring and Roots of Quaternion Polynomials

Roughly speaking, the process of factoring a quaternion polynomial is the opposite
of multiplying quaternion polynomials. Recall that usually when we factor a real
number, we are looking for prime factors that multiplied together give this number;
for example $6 = 2 \times 3$, or $12 = 2 \times 2 \times 3$. When we factor a quaternion polynomial,
it is often useful to look for simpler polynomials that multiplied together give us

the polynomial we started with. In doing so, we are usually only interested in decomposing into simple polynomials, by means of intuitive considerations. For example, the polynomial $p^3 + p(1 - j + k)p - ip^2 + (1 + j + k)p$ can be factored as a product of three simple polynomials: $(p - i)(p + i - j + k)p$. We can't do any better, so we can say the polynomial has been factored completely. Of course, an intimately related concept is that of a root of a polynomial. A quaternion number is then called a root of order $l \leq n$ of the quaternion polynomial $P_n(p)$, if $P_n(q) = (p - q)^l M_{n-l}(p)$, where $M_{n-l}(p)$ is a quaternion polynomial of degree $n - l$, and such that $M_{n-l}(q) \neq 0_{\mathbb{H}}$.

All that is needed to understand how to factor a quaternion polynomial is:

(i) *Make sure that the monomials in the polynomial are written in descending (resp. ascending) order of degrees;*

(ii) *List all the possible ways to make sure there is no common monomial that you can factor out of the polynomial.*

Example. Let $\alpha_1, \alpha_2 \in \mathbb{H}$ be roots of the quaternion polynomial $q(p) = p^2 - (1 + i + j - k)p + \lambda$, $\lambda \in \mathbb{H}$. Find λ such that $\alpha_1 + \alpha_2 = 1 - j$ and $\alpha_1^2 + \alpha_2^2 = 1 - i - j - k$.

Solution. Since $\alpha_1, \alpha_2 \in \mathbb{H}$ are both roots of $q(p)$, it follows that $\alpha_1^2 - (1 + i + j - k)\alpha_1 + \lambda = 0_{\mathbb{H}}$ and $\alpha_2^2 - (1 + i + j - k)\alpha_2 + \lambda = 0_{\mathbb{H}}$. Then a direct computation shows that $\alpha_1^2 + \alpha_2^2 - (1 + i + j - k)(\alpha_1 + \alpha_2) + 2\lambda = 0_{\mathbb{H}}$. Hence, $\lambda = \frac{1}{2} + \frac{1}{2}i + \frac{1}{2}j - \frac{1}{2}k$.

Exercise 10.247. *Let $\alpha_1, \alpha_2 \in \mathbb{H}$ be roots of the quaternion polynomial $g(p) = p^2 - pj - ip + \lambda$, $\lambda \in \mathbb{H}$. Find λ such that $\alpha_1\alpha_2 = k$, and $\alpha_1 - \alpha_2 = i - j$.*

Solution. $g(p) = p^2 - pj - ip + k$.

Exercise 10.248. *Prove that $p = i - j$ is a root of second order of the polynomial $f(p) = p^2 - pi + pj - ip + jp - 2$.*

Before proceeding, let's pause for a moment to develop some intuition about which motivation is likely to succeed. For natural numbers $n, k \geq 0$, the central binomial coefficient $\binom{n}{k}$ plays an important role in many areas of mathematics, including statistics, combinatorics and number theory. For instance, the underlying transformation techniques are used extensively in the analysis of series and in their closed-form representations. Analogous to the natural numbers, an easy example for the relationship between the central binomial coefficient and quaternions may be the following. If p is an arbitrary element of $\mathbb{H}$ and we consider the monomials $(1_{\mathbb{H}} + p)^n$ and develop them as binomials whenever $|p| < 1$, we may see that the binomial coefficient $\binom{n}{k}$ is defined as the coefficient of the monomial p^k.

It is time to explore the possibility of introducing binomial coefficients involving quaternions.

10.5   Quaternion Binomial

Let p be a real quaternion. We define the quaternion binomial coefficient as

$$\binom{p}{n} := \frac{p\,(p-1)\cdots(p-n+1)}{n!}$$

for all integers $n \geq 0$, with initial values $\binom{p}{0} = 1_{\mathbb{H}}$ and $\binom{p}{1} = p$. When n is a negative integer the binomial coefficient $\binom{p}{n}$ equals zero.

Of course, instead of writing $\sum_{k=0}^{n} \binom{p}{k}$, we can also write $\sum_{k=-\infty}^{\infty} \binom{p}{k}$, because the underlying binomial coefficients vanish on all of the extra values of k that appear in the second form of the sum. These conventions may save us a lot of work in the sequel, mainly in that we won't have to worry about changing the limits of summation if we change the variable of summation by a constant shift.

Although the best motivation for the above characterization is the fact that it works, let's pause for a moment before proceeding, to see why it is likely to work. One may show by elementary reasoning that the neutral quaternion binomial elements of addition and multiplication, and the associated results of addition (resp. subtraction) and multiplication are unique. From these conditions one can derive all the usual rules for manipulation of quaternion binomial coefficients. The student who is not thoroughly familiar with this way of introducing binomial coefficients should consult any textbook in which a full axiomatic treatment of binomial coefficients is given.

We begin by evaluating certain quaternion binomial coefficients by application of a fundamental recurrence relation in much the same manner as ordinary binomial coefficients may be treated.

10.6   Basic Relations of Quaternion Binomials

Let p be a quaternion. For any nonnegative integer n, we have the following combinatorial identities:

(i) $\quad \binom{p}{n} = \frac{p}{n}\binom{p-1}{n-1}$;

(ii) $\quad \binom{p}{n} + \binom{p}{n+1} = \binom{p+1}{n+1} \quad$ (Pascal's rule or addition formula);

(iii) $\quad \binom{p}{n}\,n = p\binom{p-1}{n-1} \quad$ (absorption property);

(iv) $\quad \binom{p}{n}\binom{n}{l} = \binom{p}{l}\binom{p-l}{n-l} \quad$ (subset-of-a-subset identity);

$$\text{(v)} \quad \sum_{l=0}^{n} \binom{p+l}{l} = \binom{p+1+n}{n};$$

$$\text{(vi)} \quad \sum_{l=0}^{n} \binom{p-l}{n-l} = \binom{p+1}{n}.$$

The first statement is immediate. Next, to prove the well-known Pascal's rule involving quaternion binomial coefficients, we use induction. For $n = 0$ the assertion is evident. Let $n \geq 1$. Since

$$\binom{p}{n} = \frac{p(p-1)\cdots(p-n+1)}{n!}$$

and

$$\binom{p}{n+1} = \frac{p(p-1)\cdots(p-n)}{(n+1)\,n!},$$

by algebraic manipulation, we deduce that

$$\binom{p}{n} + \binom{p}{n+1} = \frac{p(p-1)(p-2)\cdots(p-n+1)}{n!}\left(1 + \frac{p-n}{n+1}\right)$$

$$= \frac{p(p-1)\cdots(p-n+1)}{n!}\,\frac{p+1}{n+1}$$

$$= \binom{p+1}{n+1}.$$

For the absorption property of quaternion binomial coefficients, a straightforward algebraic manipulation shows that

$$\binom{p}{n}\,n = p\,\frac{(p-1)\cdots(p-n+1)}{(n-1)!}$$

$$= p\binom{p-1}{n-1}.$$

By algebraic calculation, we can easily deduce the well-known subset-of-a-subset identity for quaternion binomial coefficients. Now we shall prove property (v). For $n = 0$ the equality is evident. We suppose that

$$\sum_{l=0}^{n} \binom{p+l}{l} = \binom{p+1+n}{n}.$$

By Statement (ii),

$$\sum_{l=0}^{n+1}\binom{p+l}{l} = \sum_{l=0}^{n}\binom{p+l}{l} + \binom{p+1+n}{n+1}$$

$$= \binom{p+1+n}{n} + \binom{p+1+n}{n+1}$$

$$= \binom{p+2+n}{n+1}.$$

Lastly, property (vi) simply reverses the order of summation of elements in a quaternion binomial coefficient. The summation may be written as follows

$$\binom{p}{n} + \binom{p-1}{n-1} + \binom{p-2}{n-2} + \cdots + \binom{p-n+1}{1} + \binom{p-n}{0}.$$

Now, reversing the order of summation yields the sum

$$\binom{p-n}{0} + \binom{p-n+1}{1} + \cdots + \binom{p-2}{n-2} + \binom{p-1}{n-1} + \binom{p}{n},$$

which includes $n+1$ entries downward. By the previous statement, the value of the previous sum is the quaternion binomial coefficient $\binom{p+1}{n}$.

Example. Establish the following identity:

$$\binom{p-1}{n} - \binom{p-1}{n-1} = p^{-1}\binom{p}{n}(p-2n).$$

Solution. For any nonnegative integer n, we note that

$$\binom{p-1}{n} - \binom{p-1}{n-1} = \frac{(p-1)(p-2)\cdots(p-n+1)}{(n-1)!}\left(\frac{p-n}{n}-1\right)$$

$$= p^{-1}\frac{p(p-1)\cdots(p-n+1)}{n!}(p-2n).$$

Exercise 10.249. *Compute the following binomials:* (a) $\binom{i-j}{2} + \binom{i-j}{3}$; (b) $\binom{1+i+j+k}{2}$.

Solution. (a) $-\frac{1}{2}i + \frac{1}{2}j$;   (b) $-\frac{3}{2} + \frac{1}{2}i + \frac{1}{2}j + \frac{1}{2}k$.

Let us move on to the consideration of inverse quaternion binomial coefficients.

10.7   Inverse of the Quaternion Binomial

Let p be a real quaternion. We define the inverse quaternion binomial coefficient as

$$\binom{p}{n}^{-1} = n!\,(p-n+1)^{-1}\cdots(p-1)^{-1}\,p^{-1}$$

for all integers $n \geq 0$, with initial values $\binom{p}{0}^{-1} = 1_{\mathbb{H}}$ and $\binom{p}{1}^{-1} = p^{-1}$. To begin with, we can easily verify the relations

$$\binom{p}{n}\binom{p}{n}^{-1} = \binom{p}{n}^{-1}\binom{p}{n} = 1_{\mathbb{H}}$$

for any quaternion p and any nonnegative integer n. As a practical matter, when n is a negative integer the inverse binomial coefficient $\binom{p}{n}^{-1}$ equals zero. We observe that

$$\binom{p}{n}^{-1} = (n-1)!\,(p-n+1)^{-1}\cdots(p-1)^{-1}\,p^{-1}\,(p-(p-n))$$

$$= (n-1)!\,(p-n+1)^{-1}\cdots(p-1)^{-1}$$
$$- (p-n+2)\,(n-1)!\,(p-n+2)^{-1}\cdots(p-1)^{-1}\,p^{-1}\,(p-n)$$

$$= \binom{p-1}{n-1}^{-1} - (p-n+2)\binom{p}{n-1}^{-1}(p-n). \tag{10.1}$$

Exercise 10.250. *Compute* $\binom{1+j}{2}^{-1}$.

Solution. $-1 - j$.

We proceed by evaluating certain inverses of quaternion binomial coefficients by application of a fundamental recurrence relation.

10.8   Basic Relations of Quaternion Inverse Binomials

Let p be a real quaternion. For any nonnegative integer n, we have the following combinatorial identities:

(i) $\quad \dbinom{p}{n}^{-1} = n \dbinom{p-1}{n-1}^{-1} p^{-1};$

(ii) $\quad \dbinom{p}{n}^{-1} + \dbinom{p}{n+1}^{-1} = \dbinom{p+1}{n+1}^{-1} \dfrac{(p-n)^{-1}(p+1)^2}{n+1}$ $\quad$ (Pascal's rule);

(iii) $\quad \dbinom{p}{n}^{-1} \dbinom{n}{l}^{-1} = \dbinom{p-l}{n-l}^{-1} \dbinom{p}{l}^{-1}$ $\quad$ (subset-of-a-subset identity);

(iv) $\quad \displaystyle\sum_{r=0}^{n} \dbinom{p+r}{r}^{-1} = 1_{\mathbb{H}} + \left[\prod_{l=1}^{n} \dbinom{p+l}{l} \right]^{-1} \sum_{m=1}^{n} \prod_{l=1,l\neq m}^{n} \dbinom{p+l}{l};$

(v) $\quad \displaystyle\sum_{l=0}^{\infty} \dbinom{p+l}{l}^{-1} = 1_{\mathbb{H}} + (p+2)^{-1} p^{-1}, \quad$ for any $p \in \mathbb{H} \setminus \{-2_{\mathbb{H}}, 0_{\mathbb{H}}\}.$

For property (i), a straightforward algebraic manipulation yields

$$\dbinom{p}{n}^{-1} \frac{1}{n} = (n-1)! \, (p-n+1)^{-1} \cdots (p-1)^{-1} p^{-1}$$

$$= \dbinom{p-1}{n-1}^{-1} p^{-1}.$$

For the proof of the Pascal's rule we use induction. For $n = 0$ the assertion is evident. Let $n \geq 1$, and denote

$$A := \dbinom{p}{n}^{-1} + \dbinom{p}{n+1}^{-1}.$$

By algebraic manipulation, we deduce that

$$\dbinom{p}{n} A = 1_{\mathbb{H}} + \dbinom{p}{n}\dbinom{p}{n+1}^{-1} = 1_{\mathbb{H}} + (n+1)(p-n)^{-1},$$

$$\dbinom{p}{n+1} A = \dbinom{p}{n+1}\dbinom{p}{n}^{-1} + 1_{\mathbb{H}} = \frac{p-n}{n+1} + 1_{\mathbb{H}}.$$

From these two equations we obtain

$$\left[\binom{p}{n} + \binom{p}{n+1}\right] A = \frac{\left[(n+2+p)(p-n) + (n+1)^2\right](p-n)^{-1}}{n+1}$$

$$= \frac{(p-n)^{-1}(p+1)^2}{n+1},$$

whence

$$A = \binom{p+1}{n+1}^{-1} \frac{(p-n)^{-1}(p+1)^2}{n+1}.$$

By algebraic calculation, we can easily deduce the well-known subset-of-a-subset identity for the quaternion binomial coefficients. For (iv), consider

$$B = \binom{p}{0}^{-1} + \binom{p+1}{1}^{-1} + \cdots + \binom{p+n}{n}^{-1},$$

that is,

$$B - 1_{\mathbb{H}} = \binom{p+1}{1}^{-1} + \binom{p+2}{2}^{-1} + \cdots + \binom{p+n}{n}^{-1}.$$

Multiplying the left-hand side of the last equality by $\binom{p+1}{1}$, we obtain

$$\binom{p+1}{1} B - \binom{p+1}{1} - 1_{\mathbb{H}}$$

$$= \binom{p+1}{1}\left[\binom{p+2}{2}^{-1} + \binom{p+3}{3}^{-1} + \cdots + \binom{p+n}{n}^{-1}\right].$$

In a similar manner, we multiply the left-hand side of the last equality by $\binom{p+2}{2}$ and obtain

$$\binom{p+2}{2}\binom{p+1}{1} B - \binom{p+2}{2}\binom{p+1}{1} - \binom{p+2}{2} - \binom{p+1}{1}$$

$$= \binom{p+1}{1}\binom{p+2}{2}\left[\binom{p+3}{3}^{-1} + \binom{p+4}{4}^{-1} + \cdots + \binom{p+n}{n}^{-1}\right].$$

Repeating this process, we arrive at

$$\prod_{l=1}^{n}\binom{p+l}{l}B = \sum_{m=1}^{n}\prod_{l=1,l\neq m}^{n}\binom{p+l}{l} + \prod_{l=1}^{n}\binom{p+l}{l}.$$

Multiplying the left-hand side of this last equality by

$$\left[\prod_{l=1}^{n}\binom{p+l}{l}\right]^{-1}$$

we finally obtain

$$B = 1_{\mathbb{H}} + \left[\prod_{l=1}^{n}\binom{p+l}{l}\right]^{-1}\sum_{m=1}^{n}\prod_{l=1,l\neq m}^{n}\binom{p+l}{l}.$$

Consequently,

$$\sum_{r=0}^{n}\binom{p+r}{r}^{-1} = 1_{\mathbb{H}} + \left[\prod_{l=1}^{n}\binom{p+l}{l}\right]^{-1}\sum_{m=1}^{n}\prod_{l=1,l\neq m}^{n}\binom{p+l}{l}.$$

Now, we observe that

$$\sum_{l=0}^{\infty}\binom{p+l}{l}^{-1} = \binom{p+0}{0}^{-1} + \sum_{l=1}^{\infty}\binom{p+l}{l}^{-1}$$

$$= 1_{\mathbb{H}} + \sum_{l=0}^{\infty}\binom{p+l+1}{l+1}^{-1}.$$

Let $p \in \mathbb{H} \setminus \{-2_{\mathbb{H}}, 0_{\mathbb{H}}\}$. Applying (10.1) to each term of the sums, we have

$$\sum_{l=0}^{\infty}\binom{p+l}{l}^{-1} = 1_{\mathbb{H}} + \sum_{l=0}^{\infty}\left[\binom{p+l}{l}^{-1} - (p+2)\binom{p+l+1}{l}^{-1}p\right],$$

and, so

$$(p+2)\sum_{l=0}^{\infty}\binom{p+l+1}{l}^{-1}p = 1_{\mathbb{H}}.$$

Hence, we obtain

$$\sum_{l=0}^{\infty} \binom{p+l}{l}^{-1} = 1_{\mathbb{H}} + (p+2)^{-1} p^{-1}.$$

completing our proof.

Example. Prove the following identity:

$$\binom{p-1}{n}^{-1} - \binom{p-1}{n-1}^{-1} = -\frac{p\,(p-2n)}{n} (p-n)^{-1} \binom{p}{n}^{-1}.$$

Solution. First observe that

$$\binom{p-1}{n} - \binom{p-1}{n-1} = (p-2n)\, p^{-1} \binom{p}{n}.$$

Denoting

$$C := \binom{p-1}{n}^{-1} - \binom{p-1}{n-1}^{-1}$$

we have the relations

$$\binom{p-1}{n} C = 1_{\mathbb{H}} - \frac{p-n}{n},$$

$$\binom{p-1}{n-1} C = n(p-n)^{-1} - 1_{\mathbb{H}}.$$

Hence by an argument such as was just used above, we finally obtain

$$\binom{p-1}{n}^{-1} - \binom{p-1}{n-1}^{-1} = -\frac{p\,(p-2n)}{n} (p-n)^{-1} \binom{p}{n}^{-1}.$$

Exercise 10.251. *Compute the following quaternion inverse binomials:* (a) $\binom{p}{0}^{-1} + \binom{p}{3}^{-1} + \binom{p}{6}^{-1} + \ldots;$ (b) $\binom{p}{1}^{-1} + \binom{p}{4}^{-1} + \binom{p}{7}^{-1} + \cdots;$ (c) $\binom{p}{2}^{-1} + \binom{p}{5}^{-1} + \binom{p}{8}^{-1} + \cdots;$ (d) $\binom{p}{0}^{-1} + \binom{p}{4}^{-1} + \binom{p}{8}^{-1} + \cdots;$ (e) $\binom{p}{1}^{-1} + \binom{p}{5}^{-1} + \binom{p}{9}^{-1} + \cdots;$ (f) $\binom{p}{2}^{-1} + \binom{p}{6}^{-1} + \binom{p}{10}^{-1} + \cdots;$ (g) $\binom{p}{3}^{-1} + \binom{p}{7}^{-1} + \binom{p}{11}^{-1} + \cdots.$

Solution. (a) $\frac{1}{3}(2^p + 2\cos\frac{p\pi}{3})$;   (b) $\frac{1}{3}(2^p + 2\cos\frac{(p-2)\pi}{3})$;   (c) $\frac{1}{3}(2^p + 2\cos\frac{(p-4)\pi}{3})$;   (d) $\frac{1}{2}(2^{p-1} + 2^{\frac{p}{2}}\cos\frac{p\pi}{4})$;   (e) $\frac{1}{2}(2^{p-1} + 2^{\frac{p}{2}}\sin\frac{p\pi}{4})$;   (f) $\frac{1}{2}(2^{p-1} - 2^{\frac{p}{2}}\cos\frac{p\pi}{4})$;   (g) $\frac{1}{2}(2^{p-1} - 2^{\frac{p}{2}}\sin\frac{p\pi}{4})$.

Example. For $p \neq 0_{\mathbb{H}}$, define $J_n = \sum_{l=0}^{\infty}(-1)^l \binom{p+l}{l}^{-1}$. Prove that J_n satisfies the recursion relation $J_{n+1} = (p+2)^{-1}\,(2J_n - 1_{\mathbb{H}})\,p^{-1}$.

Solution. For $p \neq 0_{\mathbb{H}}$, a first computation shows that

$$J_n = \sum_{l=0}^{\infty}(-1)^l\binom{p+l}{l}^{-1}$$

$$= (-1)^0\binom{p+0}{0}^{-1} + \sum_{l=0}^{\infty}(-1)^{l+1}\binom{p+(l+1)}{l+1}^{-1}.$$

Applying the identity (10.1) to each term of the sums we have

$$J_n = 1_{\mathbb{H}} + \sum_{l=0}^{\infty}(-1)^{l+1}\left[\binom{p+l}{l}^{-1} - (p+2)\binom{p+l+1}{l}^{-1}p\right]$$

$$= 1_{\mathbb{H}} - \sum_{l=0}^{\infty}(-1)^l\binom{p+l}{l}^{-1} + (p+2)\sum_{l=0}^{\infty}(-1)^l\binom{p+l+1}{l}^{-1}p$$

$$= 1_{\mathbb{H}} - J_n + (p+2)\,J_{n+1}\,p\,,$$

and the recursion relation follows.

Exercise 10.252. *Compute the following binomials:* (a) $\sum_{l=0}^{n}\binom{p}{l}^{-1}\binom{q}{n-l}^{-1}$;   (b) $\sum_{l=0}^{n}\binom{p}{l}^{-1}\binom{q-p}{n-l}^{-1}$.

Solution. (a) $n!\sum_{l=0}^{n}((p-l)!)^{-1}((q-n+l)!)^{-1}$;   (b) $n!\sum_{l=0}^{n}((p-l)!)^{-1}((q-p-n+l)!)^{-1}$.

10.9   Advanced Practical Exercises

1. Simplify: (a) $iqjq(i+j)q + iqj + 2iqjq(i+j)q - 3iqj$;   (b) $2iqjq(i-j)qj + qjqi + 2iqj$.

2. Find the degree of the following monomials: (a) $iqjqi$;   (b) $qiq(1+i+j)qi$;   (c) $jqiq(1-i-j+k)qiqj$.

3. Compute: (a) $(iqj+1+i)^2$; (b) $((1+i)q(1-j)qk+iq)(kqk+jqjqj+k)$; (c) $(-iqjqj-kqi)(iqj+jqk+i)$.

4. Find $f(j)$ where: (a) $f(q) = (iqkqj - jqk)(iqjqi + iqi + jqk)$; (b) $f(q) = (iqkqj + iqi)^2$.

5. Check the validity of the following equalities: (a) $iqj+qj+iq = (2i+1)q(j+1)$; (b) $iqj+qj+iq = (i+1)qj+iq$; (c) $(1+i-j)qjqi+jqjqi+iqj = (1+i)qjqi+iqj$; (d) $qj+iqj+jq = (1+i+j)q(2j+1)$.

6. Let $a \sim b$, $a, b, c \in \mathbb{H}\setminus\{0_\mathbb{H}\}$ and $acb^{-1} = c$. Prove that the equation $ax + xb = c$ has a non-zero solution.

7. Find a solution $x \in \mathbb{H}$ to the following equations: (a) $x + xj = k$; (b) $(1+i)x + x(1+i+j+k) = i$; (c) $kx + xj = i$; (d) $ix + xj = k$; (e) $jx + xk = i$.

8. Find a solution to the following equations: (a) $ix + ixj + xk = 1+i-j+k$; (b) $(1+j)x - kxj + xi = i+j$; (c) $kx + kxi + xj = i+j+k$.

9. Solve the systems

$$\text{(a)} \begin{cases} ix + yj = 1+i+j, \\ xk + yi = 1-i-j-k, \end{cases} ; \qquad \text{(b)} \begin{cases} (1+i)x + yk = 1-j, \\ jx + yk = i. \end{cases}$$

10. Solve the systems

$$\text{(a)} \begin{cases} (1+j)x + yi = 1+i-k, \\ kx + iyj + yk = i+j+k, \end{cases} ; \qquad \text{(b)} \begin{cases} (1-k)x + jyk = i, \\ jx + kyj + yi = 1-i-j. \end{cases}$$

11. Solve the system

$$\begin{cases} (1+i)x + yj + izk = 1-i-k, \\ (1-i)x + zj = 1+i+j+k, \\ ixj + zk - iyj = i+k. \end{cases}$$

12. Factor the following polynomials: (a) $p^2 - 2q^2 + 2qp - pq$; (b) $2p^2 + q^2 + qp + 2pq$; (c) $qp^2 - 2q^3 + q^2p - 2qpq$; (d) $p^3 - 2pqp - 3qp^2 + 6q^2p$; (e) $p^3 - q^3 + p^2q - q^2p - qp^2 + pqp + pq^2 - qpq$; (f) $2p^3 + q^3 + p^2q - pq^2 - 2qp^2 + qpq - 2q^2p$; (g) $2p^4 + 6q^4 + 4p^3q - 3qp^3 - 2p^2qp - 4p^2q^2 + 2pqp^2 + 4pqpq - 2pq^2p - 4pq^3 - 6qp^2q + 3qpqp + 6qpq^2 - 3q^2p^2 - 6q^2pq + 3q^3p$; (h) $2p^3 - 3q^3 + 6p^2q - 3pq^2 + 2qp^2 - pqp + 6qpq - q^2p$; (i) $p^2qp^3q^2 + p^4qp^2q$; (j) $pqp + p^2qp^3$.

13. Let $\alpha_1, \alpha_2 \in \mathbb{H}$ be roots of the polynomial $q(p) = p^2 + (1 + i + j - k)p + \mu$, $\mu \in \mathbb{H}$. Find μ such that $\alpha_1\alpha_2 = i - j$, $\alpha_2\alpha_1 = 1 + k$, and $q(\alpha_1 + \alpha_2) = 1 - i - j - k$.

14. Let $\alpha_1, \alpha_2 \in \mathbb{H}$ be roots of the polynomial $q(p) = p^2 + \lambda p + \mu$, $\lambda, \mu \in \mathbb{H}$. Find λ and μ so that $\alpha_1\alpha_2 = \alpha_2\alpha_1$, $\alpha_1 \neq \alpha_2$, $\alpha_1 + \alpha_2 = 1 - i - j - k$, and $q(\alpha_1 + \alpha_2) = j + k$.

15. Let $\alpha_1, \alpha_2 \in \mathbb{H}$ be roots of the polynomial $q(p) = p^2 + \lambda p + 2\lambda$, with $\lambda \in \mathbb{H}$. Find λ such that $\alpha_1 + \alpha_2 = 1 - k$ and $\alpha_1^2 + \alpha_2^2 = i + j + k$.

16. Find a polynomial $q(p) = ipa + p^2b - p$, with $a, b \in \mathbb{H}$, such that $q(i) = -1 - 2i - 2j - k$ and $q(k) = -i - j - 2k$.

17. Find a polynomial $q(p) = ipbpk - p^2 - ap - kp + 1 + i + j + k$, $a, b \in \mathbb{H}$ such that $q(i) = 2 + i + k$ and $q(j) = 3 + 2i + j$.

18. Prove that $p = j$ is a root of order 3 of the polynomial $f(p) = p^3 - p^2j - jp^2 - pjp + jpj - 2p + j$.

19. Find the parameter $\alpha \in \mathbb{H}$ such that $p = -i + k$ is a root of order 3 of the polynomial

$$f(p) = p^3 + p^2(i - k) + (i - k)p^2 + p(i - k)p + (i - k)p(i - k) - 4p + \alpha.$$

20. Find the parameter $\alpha \in \mathbb{H}$ such that $p = -i + k$ is a root of order 3 of the polynomial

$$f(p) = p^3 + p^2\alpha + (i - k)p^2 + p(i - k)p + (i - k)p(i - k) - 4p - 2i + 2k.$$

21. Let p be a real quaternion. Show that $\binom{p+m}{m}^{-1} = m \binom{p+m}{m-1}^{-1}(p + 1)^{-1}$, for any integer m.

22. Compute the following quaternion inverse binomials: (a) $\sum_{l=0}^{2n} l \binom{p}{l}^{-1}$; (b) $\sum_{l=0}^{2n}(-1)^l \binom{p}{l}^{-1}\binom{p}{2n-l}^{-1}$; (c) $\sum_{l=0}^{n} l \binom{p}{l}^{-2}$.

Solutions

11

Chapter 1

1. (a) $p + q = 3 + i + 5i + j - 2k = 3 + 6i + j - 2k$;   (b) $3 - 4i - j + 2k$;
   (c) $6 - 33i - 7j + 14k$;   (d) $9 + 23i + 4j - 8k$.
2. (a) -1;   (b) 1;   (c) 1;   (d) -1;   (e) -1;   (f) 1.
3.

$$(a) \quad pq = (1 + i - j - k)(3 + 2i + j + 4k)$$

$$= 3 + 2i + j + 4k + 3i - 2 + k - 4j - 3j + 2k + 1$$

$$- 4i - 3k - 2j + i + 4 = 6 + 2i - 8j + 4k;$$

   (b) $3 + 3i - 3j + k$;   (c) $-1 + 4i + 7j + 12k$;   (d) $14 - 2i - 8j + 24k$;
   (e) $-2 + 2i - 2j - 2k$;   (f) $-162 + 12i + 6j + 24k$;   (g) $14 - 22i + 24j + 68k$;
   (h) $-12 + 24i - 26j - 70k$;   (i) $1 + 4i - j + 2k$;   (j) $-147 - 9i + 29j + 91k$.
4. (a) First we consider $p(1 + i - j) = 1_{\mathbb{H}}$. We have

$$1_{\mathbb{H}} = (a + bi + cj + dk)(1 + i - j)$$
$$= (a - b + c) + (a + b + d)i + (-a + c + d)j + (-b - c + d)k.$$

From here we obtain the following system for a, b, c and d:

$$\begin{cases} a - b + c = 1, \\ a + b + d = 0, \\ -a + c + d = 0, \\ -b - c + d = 0. \end{cases}$$

J.P. Morais et al., *Real Quaternionic Calculus Handbook*,
DOI 10.1007/978-3-0348-0622-0_11, © Springer Basel 2014

Its solution is

$$a = \frac{1}{3}, \ b = -\frac{1}{3}, \ c = \frac{1}{3}, \ d = 0.$$

Consequently,

$$p = \frac{1}{3} - \frac{1}{3}i + \frac{1}{3}j.$$

It is easy to check that

$$\left(\frac{1}{3} - \frac{1}{3}i + \frac{1}{3}j\right)(1 + i - j) = 1_{\mathbb{H}}.$$

Now we consider

$$(1 + i - j)p = 1_{\mathbb{H}}.$$

We have

$$1_{\mathbb{H}} = (1 + i - j)(a + bi + cj + dk)$$
$$= a - b + c + (b + a - d)i + (c - a - d)j + (d + c + b)k.$$

From the last equality we get for a, b, c, d the system

$$\begin{cases} a - b + c = 1, \\ a + b - d = 0, \\ c - a - d = 0, \\ d + c + b = 0. \end{cases}$$

Its solution is

$$a = \frac{1}{3}, \ b = -\frac{1}{3}, \ c = \frac{1}{3}, \ d = 0.$$

Therefore,

$$p = \frac{1}{3} - \frac{1}{3}i + \frac{1}{3}j.$$

It is easy to check that

$$(1 + i - j)\left(\frac{1}{3} - \frac{1}{3}i + \frac{1}{3}j\right) = 1_{\mathbb{H}}.$$

(b) $p = -i$;   (c) $p = -j$;   (d) $p = -k$.

5. (a), (b)

$$p = \frac{a}{a^2 + b^2 + c^2 + d^2} - \frac{b}{a^2 + b^2 + c^2 + d^2}i$$

$$- \frac{c}{a^2 + b^2 + c^2 + d^2}j - \frac{d}{a^2 + b^2 + c^2 + d^2}k.$$

6. (a) We have

$$2p = 1 - i + j + 3k,$$

which implies

$$2a + 2bi + 2cj + 2dk = 1 - i + j + 3k.$$

From here we obtain for a, b, c, d the following system

$$\{2a = 1, \ 2b = -1, \ 2c = 1, \ 2d = 3.$$

Consequently,

$$p = \frac{1}{2} - \frac{1}{2}i + \frac{1}{2}j + \frac{3}{2}k.$$

It is easy to check that p satisfies the given equation. (b) $p = \frac{1 \pm \sqrt{5}}{2}$;   (c) $p = a + \frac{1}{2a-1}i$, where a is a solution to the equation

$$4a^4 - 8a^3 + 5a^2 - a - 1 = 0 \tag{11.1}$$

(such solution exists: prove!), $a \neq \frac{1}{2}$; (d) $p = a + \frac{1}{2a-1}j$;   (e) $p = a + \frac{1}{2a-1}k$, where a is a solution to the Eq. (11.1); (f) $p = \frac{1}{2} + \frac{1}{2}i + k$;   (g) $p = \frac{1}{2} + \frac{1}{2}i + j$; (h) $p = -\frac{3}{7} + \frac{5}{7}i - \frac{5}{7}j + \frac{5}{7}k$;   (i) $p = bi + cj + dk$ so that $b^2 + c^2 + d^2 = 1$; (j) $p = \pm\frac{\sqrt{3}}{2} - \frac{1}{2}i$;   (k) $p = \pm\frac{1}{2} - \frac{1}{2}i \pm \frac{1}{2}j - \frac{1}{2}k$;   (l) $p = \pm\frac{1}{2} - \frac{1}{2}i + \frac{1}{2}j \pm \frac{1}{2}k$.

7. (a) $2 + i - j$;   (b) $1 - i - j - k$;   (c) $2 - 2i - 4j$;   (d) $2 - 2j - 4k$; (e) $2 - 2i - 4j$;   (f) $2 - 2j - 4k$.

8.

(a) $(a_1 a_2 - b_1 b_2 - c_1 c_2 - d_1 d_2) - (a_1 b_2 + b_1 a_2 + c_1 d_2 - d_1 c_2)i$

$- (a_1 c_2 - b_1 d_2 + c_1 a_2 + d_1 b_2)j - (a_1 d_2 + b_1 c_2 - c_1 b_2 + d_1 a_2)k;$

(b) $(a_1 a_2 - b_1 b_2 - c_1 c_2 - d_1 d_2) + (-a_1 b_2 - b_1 a_2 + c_1 d_2 - d_1 c_2)i$

$+ (-a_1 c_2 - b_1 d_2 - c_1 a_2 + d_1 b_2)j + (-a_1 d_2 + b_1 c_2 - c_1 b_2 - d_1 a_2)k;$

(c) $(a_1a_2 - b_1b_2 - c_1c_2 - d_1d_2) - (a_2b_1 + a_1b_2 + c_2d_1 - d_2c_1)\,i$

$\quad - (a_2c_1 - b_2d_1 + c_2a_1 + d_2b_1)\,j - (a_2d_1 + b_2c_1 - c_2b_1 + d_2a_1)\,k;$

(d) $(a_1a_2 - b_1b_2 - c_1c_2d_1d_2) - (a_2b_1 + b_2a_1 + c_2d_1 - d_2c_1)\,i$

$\quad - (a_2c_1 - b_2d_1 + c_2a_1 + d_2b_1)\,j - (a_2d_1 + b_2c_1 - c_2b_1 + d_2a_1)\,k.$

9. **Hint.** Use the previous exercise.

10. We have that $pp^{-1} = 1_{\mathbb{H}}$ and $|p|^2 = p\overline{p}$. Consequently,

$$\frac{\overline{p}}{|p|^2} = \frac{\overline{p}}{p\overline{p}} = p^{-1}.$$

11. (a) Here

$$a = 1, \quad b = -2, \quad c = 3, \quad d = 4.$$

Then the first matrix representation is

$$\begin{pmatrix} 1 - 4i & 2 + 3i \\ -2 + 3i & 1 + 4i \end{pmatrix};$$

(b) $\begin{pmatrix} 2 & 1+i \\ -1+i & 2 \end{pmatrix};$ (c) $\begin{pmatrix} \frac{1}{2} - i & -\frac{1}{3} - 2i \\ \frac{1}{3} - 2i & \frac{1}{2} + i \end{pmatrix}.$

12. (a) Let $p = a + bi + cj + dk$ $(a, b, c, d \in \mathbb{R})$ be a quaternion that corresponds to the given matrix. Then

$$\begin{pmatrix} 1 - i & 2 + 4i \\ -2 + 4i & 1 + i \end{pmatrix} = \begin{pmatrix} a - di & -b + ci \\ b + ci & a + di \end{pmatrix},$$

therefore we obtain the system

$$\begin{cases} a - di = 1 - i, \\ -b + ci = 2 + 4i. \end{cases}$$

Consequently, $a = 1, b = -2, c = 4$, and $d = 1$. That is,

$$p = 1 - 2i + 4j + k;$$

(b) $p = 1 + 2i + j;$ (c) $p = 1 - 2i + 8j - \frac{1}{2}k.$

13. (a) Let us suppose that the given matrix corresponds to a quaternion

$$p = a + bi + cj + dk, \quad a, b, c, d \in \mathbb{R}.$$

Then

$$\begin{pmatrix} 1+i & 1-i \\ 1+i & 2+i \end{pmatrix} = \begin{pmatrix} a-di & -b+ci \\ b+ci & a+di \end{pmatrix},$$

whence

$$\begin{cases} a-di = 1+i, \\ -b+ci = 1-i, \\ b+ci = 1+i, \\ a+di = 2+i, \end{cases}$$

which is a contradiction. Therefore the given matrix does not correspond to any quaternion; (b) no solution;   (c) no solution;   (d) $p = 1 - 2i - j + 2k$.

14. We have that

$$|z| = \sqrt{a^2 + b^2 + c^2 + d^2}.$$

Also

$$\det A = \begin{vmatrix} a-di & -b+ci \\ b+ci & a+di \end{vmatrix} = (a-di)(a+di) - (-b+ci)(b+ci)$$

$$= a^2 + b^2 + c^2 + d^2,$$

from here $|z| = \sqrt{\det A}$.

15. (a) The first matrix representation of $p = 1 - i + 2j - k$ is

$$A_1 = \begin{pmatrix} 1+i & 1+2i \\ -1+2i & 1-i \end{pmatrix}.$$

Next,

$$A_2 = \begin{pmatrix} -i & 2+3i \\ -2+3i & i \end{pmatrix}$$

is the first matrix representation of the quaternion $q = -2i + 3j + k$. Then

$$A_1 A_2 = \begin{pmatrix} 1+i & 1+2i \\ -1+2i & 1-i \end{pmatrix} \begin{pmatrix} -i & 2+3i \\ -2+3i & i. \end{pmatrix}$$

$$= \begin{pmatrix} -7-2i & -3+6i \\ 3+6i & -7+2i \end{pmatrix}.$$

The last matrix corresponds to the quaternion

$$pq = -7 + 3i + 6j + 2k;$$

(b) $pq = 3 + 3k$.

16. $-3 + 5i - 2j + 3k$.

17. (a) Here

$$a = 1, \quad b = -2, \quad c = \frac{1}{2}, \quad d = 1.$$

Then the second matrix representation is

$$\begin{pmatrix} 1 & 2 & 1 & -\frac{1}{2} \\ -2 & 1 & -\frac{1}{2} & -1 \\ -1 & \frac{1}{2} & 1 & 2 \\ \frac{1}{2} & 1 & -2 & 1 \end{pmatrix};$$

(b) $\begin{pmatrix} 0 & -1 & 3 & 1 \\ 1 & 0 & 1 & -3 \\ -3 & -1 & 0 & -1 \\ -1 & 3 & 1 & 0 \end{pmatrix};$ (c) $\begin{pmatrix} 2 & -4 & -1 & 1 \\ 4 & 2 & 1 & 1 \\ 1 & -1 & 2 & -4 \\ -1 & -1 & 4 & 2 \end{pmatrix};$

(d) $\begin{pmatrix} 0 & -1 & 1 & -1 \\ 1 & 0 & -1 & -1 \\ -1 & 1 & 0 & -1 \\ 1 & 1 & 1 & 0 \end{pmatrix}.$

18. For $p = 1 - i + j - k$ we have the representation

$$A_1 = \begin{pmatrix} 1 & 1 & 1 & -1 \\ -1 & 1 & -1 & -1 \\ -1 & 1 & 1 & 1 \\ 1 & 1 & -1 & 1 \end{pmatrix},$$

while for $q = 2 - i - j - k$ we have

$$A_2 = \begin{pmatrix} 2 & 1 & -1 & 1 \\ -1 & 2 & 1 & 1 \\ 1 & -1 & 2 & 1 \\ -1 & -1 & -1 & 2 \end{pmatrix}.$$

Then

$$A_1 A_2 = \begin{pmatrix} 3 & 3 & 3 & 1 \\ -3 & 3 & 1 & -3 \\ -3 & -1 & 3 & 3 \\ -1 & 3 & -3 & 3 \end{pmatrix} = \begin{pmatrix} a & -b & d & -c \\ b & a & -c & -d \\ -d & c & a & -b \\ c & d & b & a \end{pmatrix},$$

which yields

$$a = 3, \quad b = -3, \quad c = -1, \quad d = 3,$$

i.e. $pq = 3 - 3i - j + 3k$.

19. (a) Let $p = a + bi + cj + dk$, with $a,b,c,d \in \mathbb{R}$ be the quaternion that corresponds to the given matrix. For its second matrix representation we have

$$\begin{pmatrix} a & -b & d & -c \\ b & a & -c & -d \\ -d & c & a & -b \\ c & d & b & a \end{pmatrix}.$$

It follows that

$$a = \frac{1}{2}, \quad b = -2, \quad c = -3, \quad d = -9,$$

i.e.

$$p = \frac{1}{2} - 2i - 3j - 9k;$$

(b) $p = 7 + i + 9k$; (c) $p = 1 - i + j + k$; (d) $p = 2 - 2i + j - k$;
(e) $p = 3 - \frac{1}{2}i + j + k$; (f) $p = 1 - \frac{1}{3}i + \frac{2}{3}j + 4k$.

20. (a) no solution; (b) no solution; (c) has a solution; (d) no solution;
(e) no solution; (f) no solution.

21. $2 - 5i + 5j + 5k$.

22. (a) Here $p_0 = 0$, $\mathbf{p} = (1,0,0)$, $q_0 = 0$, $\mathbf{q} = (1,0,0)$. Then

$$p \cdot q = p_0 q_0 + \mathbf{pq} = 0 + (1,0,0)(1,0,0) = 1;$$

(b) 1; (c) 1; (d) 0; (e) 0; (f) 0; (g) 0; (h) 0; (i) 0; (j) $-\frac{2}{3}$;
(k) -3; (l) 1.

23. (a) We have

$$pq = (1 - i - j - k)(2 + i + j) = 4 - 2j - 2k,$$

$$p_0 = 1, \quad q_0 = 2,$$

$$\mathbf{p} = (-1,-1,-1),$$

$$\mathbf{q} = (1,1,0),$$

$$p \cdot q = p_0 q_0 + \mathbf{pq} = 2 + (-1,-1,-1)(1,1,0) = 0,$$

$$p \cdot q + 2q = 4 + 2i + 2j,$$

$$p(p \cdot q + 2q) = (1 - i - j - k)(4 + 2i + 2j) = 8 - 4j - 4k,$$

$$p - q = -1 - 2i - 2j - k,$$

$$\mathrm{Sc}(p - q) = -1, \quad \mathrm{Vec}(p - q) = (-2,-2,-1),$$

$$q \cdot (p - q) = q_0 \mathrm{Sc}(p - q) + \mathbf{q}\mathrm{Vec}(p - q)$$

$$= -2 + (1,1,0)(-2,-2,-1) = -6.$$

Therefore

$$pq - p(p \cdot q + 2q) + q \cdot (p - q) = 4 - 2j - 2k - (8 - 4j - 4k) - 6$$

$$= -10 + 2j + 2k;$$

(b) -18;   (c) $-10 - 2j - 2k$.

24. We have

$$p_0 = a_1, \quad q_0 = a_2,$$

$$\mathbf{p} = (b_1, c_1, d_1), \quad \mathbf{q} = (b_2, c_2, d_2),$$

$$p \cdot q = p_0 q_0 + \mathbf{pq} = a_1 a_2 + b_1 b_2 + c_1 c_2 + d_1 d_2.$$

25. We have

$$\overline{p} = a_1 - b_1 i - c_1 j - d_1 k,$$

$$\overline{p}q = (a_1 - b_1 i - c_1 j - d_1 k)(a_2 + b_2 i + c_2 j + d_2 k)$$

$$= a_1 a_2 + b_1 b_2 + c_1 c_2 + d_1 d_2 + (a_1 b_2 - b_1 a_2 - c_1 d_2 + d_1 c_2)i$$

$$+ (a_1 c_2 + b_1 d_2 - c_1 a_2 - d_1 b_2)j + (a_1 d_2 - b_1 c_2 + c_1 b_2 - d_1 a_2)k,$$

$$\overline{q} = a_2 - b_2 i = c_2 j - d_2 k,$$

and, moreover

$$\overline{q}p = (a_2 - b_2 i - c_2 j - d_2 k)(a_1 + b_1 i + c_1 j + d_1 k)$$

$$= a_1 a_2 + b_1 b_2 + c_1 c_2 + d_1 d_2 + (a_2 b_1 - b_2 a_1 - c_2 d_1 + d_2 c_1)i$$

$$+ (a_2c_1 + b_2d_1 - c_2a_1 - d_2b_1)j + (a_2d_1 - b_2c_1 + c_2b_1 - d_2a_1)k,$$

$$\frac{1}{2}(\overline{p}q + \overline{q}p) = a_1a_2 + b_1b_2 + c_1c_2 + d_1d_2.$$

From here and from the previous exercise we obtain our equality.

26. (a) We have

$$\overline{p} = 1 + i - j,$$

$$\overline{p}q = (1 + i - j)(i - j - k) = -2 + 2i - k,$$

$$\overline{q} = -i + j + k,$$

$$\overline{q}p = (-i + j + k)(1 - i + j) = -2 - 2i + k,$$

$$p \cdot q = \frac{1}{2}(\overline{p}q + \overline{q}p) = -2;$$

(b) 0;   (c) -2.

27. It is easy to find

$$Sc(1_{\mathbb{H}}) = 1, \quad p_0 = a,$$

$$Vec(1) = (0,0,0), \quad \mathbf{p} = (b,c,d),$$

$$1_{\mathbb{H}} \cdot p = Sc(1_{\mathbb{H}})p_0 + (0,0,0)(b,c,d) = a = p_0.$$

28. (a) Here $p_0 = 1, q_0 = 0, \mathbf{p} = (-1,0,0), \mathbf{q} = (0,1,1)$. Then

$$(p,q) = p_0\mathbf{q} - q_0\mathbf{p} - \mathbf{p} \times \mathbf{q}$$

$$= (0,1,1) - (0,1,-1) = (0,0,2),$$

i.e. $(p,q) = 2k$; (b) $-3i - j - k$;   (c) $-i - j + k$.

29. (a) We have

$$p_0 = 1, \quad q_0 = 1, \quad \mathbf{p} = (-1,-1,-1), \quad \mathbf{q} = (1,1,-1),$$

$$p \cdot q = p_0q_0 + \mathbf{p}\mathbf{q} = 0,$$

$$-2p + q = -1 + 3i + 3j + k, \quad pq = 2 + 2i - 2j - 2k,$$

$$(p,q) = p_0\mathbf{q} - q_0\mathbf{p} - \mathbf{p} \times \mathbf{q}$$

$$= (1,1,-1) - (-1,-1,-1) - (2,-2,0) = (0,4,0) = 4j,$$

and also

$$\mathrm{Sc}(p,q) = 0, \quad \mathrm{Vec}(p,q) = (0,4,0),$$

$$q \cdot (p,q) = q_0 \mathrm{Sc}(p,q) + \mathbf{q}\mathrm{Vec}(p,q) = 4.$$

Consequently

$$p \cdot q - 2p + q - pq + q \cdot (p,q) = 1 + i + 5j + 3k;$$

(b) $4 - 2i - 6j + 4k$.

30. We have

$$p_0 = a_1, \quad q_0 = a_2,$$

$$Vec(p) = (b_1, c_1, d_1), \quad \mathbf{q} = (b_2, c_2, d_2),$$

$$(p,q) = p_0\mathbf{q} - q\mathbf{p} - \mathbf{p} \times \mathbf{q} =$$

$$= a_1(b_2, c_2, d_2) - a_2(b_1, c_1, d_1) - (b_1, c_1, d_1) \times (b_2, c_2, d_2)$$
$$= (a_1b_2, a_1c_2, a_1d_2) - (a_2b_1, a_2c_1, a_2d_1)$$
$$- (c_1d_2 - c_2d_1, b_2d_1 - b_1d_2, b_1c_2 - b_2c_1)$$
$$= (a_1b_2 - a_2b_1 - c_1d_2 + c_2d_1, a_1c_2 - a_2c_1 - b_2d_1 + b_1d_2,$$
$$a_1d_2 - a_2d_1 - b_1c_2 + b_2c_1),$$

i.e.

$$\mathbf{p} = (b_1, c_1, d_1), \quad \mathbf{q} = (b_2, c_2, d_2),$$

$$(p,q) = (a_1b_2 - a_2b_1 - c_1d_2 + c_2d_1)i$$
$$+ (a_1c_2 - a_2c_1 - b_2d_1 + b_1d_2)j + (a_1d_2 - a_2d_1 - b_1c_2 + b_2c_1)k.$$

31. We have

$$\overline{p} = a_1 - b_1i - c_1j - d_1k,$$

$$\overline{p}q = a_1a_2 + b_1b_2 + c_1c_2 + d_1d_2 + (a_1b_2 - b_1a_2 - c_1d_2 + d_1c_2)i +$$
$$+ (a_1c_2 + b_1d_2 - c_1a_2 - d_1b_2)j + (a_1d_2 - b_1c_2 + c_1b_2 - d_1a_2)k,$$

$$\overline{q} = a_2 - b_2i - c_2j - d_2k,$$

$$\overline{q}p = a_1a_2 + b_1b_2 + c_1c_2 + d_1d_2 + (a_2b_1 - b_2a_1 - c_2d_1 + d_2c_1)i +$$
$$+ (a_2c_1 + b_2d_1 - c_2a_1 - d_2b_1)j + (a_2d_1 - b_2c_1 + c_2b_1 - d_2a_1)k,$$

$$\frac{1}{2}\left(\overline{p}q - \overline{q}p\right) = (a_1b_2 - b_1a_2 - c_1d_2 + c_2d_1)i +$$
$$+ (a_1c_2 - a_2c_1 + b_1d_2 - b_2d_1)j + (a_1d_2 - a_2d_1 - b_1c_2 + c_1b_2)k.$$

From the previous exercise we obtain

$$(p,q) = \frac{\overline{p}q - \overline{q}p}{2}.$$

32. (a) We have

$$\overline{p} = -i - j,$$

$$\overline{p}q = (-i - j)(i + j + k) = 2 - i + j,$$

$$\overline{q} = -i - j - k,$$

$$\overline{q}p = (-i - j - k)(i + j) = 2 + i - j,$$

$$(p,q) = \frac{1}{2}\left(\overline{p}q - \overline{q}p\right) = -i + j;$$

(b) $i + 6j + 4k$;   (c) $-3i - 10j + 8k$;   (d) $3i - j + 2k$;   (e) $2i + j + 4k$;
(f) $3j$;   (g) k;   (h) j;   (i) $-i$.

33. One has that

$$(p,q) = \frac{1}{2}\left(\overline{p}q - \overline{q}p\right),$$

$$(q,p) = \frac{1}{2}\left(\overline{q}p - \overline{p}q\right) = -\frac{1}{2}\left(\overline{p}q - \overline{q}p\right) = -(p,q).$$

34. Let $p = a_1 + b_1i + c_1j + d_1k$ and $q = a_2 + b_2i + c_2j + d_2k$, where $a_i, b_i, c_i, d_i \in \mathbb{R}$ $(i = 1, 2)$. Then we have

$$\overline{p}q = a_1a_2 + b_1b_2 + c_1c_2 + d_1d_2 + (a_1b_2 - b_1a_2 - c_1d_2 + d_1c_2)i +$$
$$+ (a_1c_2 + b_1d_2 - c_1a_2 - d_1b_2)j + (a_1d_2 - b_1c_2 + c_1b_2 - d_1a_2)k,$$

and

$$\mathrm{Vec}(\overline{p},q) = (a_1b_2 - b_1a_2 - c_1d_2 + d_1c_2,$$
$$a_1c_2 + b_1d_2 - c_1a_2 - d_1b_2, a_1d_2 - b_1c_2 + c_1b_2 - d_1a_2).$$

Therefore, $(p,q) = \mathrm{Vec}(\overline{p}q)$. Also $(p,q) = -(q,p) = -\mathrm{Vec}(\overline{q}p)$.

35. (a) We have

$$p_0 = 2, \quad q_0 = 1, \quad \mathbf{p} = (-1, 1, -1), \quad \mathbf{q} = (1, 1, 1),$$

$$[p, q] = p_0 q_0 - \mathbf{pq} + p_0\mathbf{q} + q_0\mathbf{p}$$
$$= 2 - (-1, 1, -1)(1, 1, 1) + 2(1, 1, 1) + (-1, 1, -1)$$
$$= 3 + (1, 3, 1) = 3 + i + 3j + k;$$

(b) $6 + 2j + 3k$;   (c) $-2 - i + 4j + 2k$;   (d) $0_{\mathbb{H}}$;   (e) $0_{\mathbb{H}}$;   (f) $0_{\mathbb{H}}$.

36. We immediately get

$$2p = 2 + 2i + 2j - 2k, \quad 2p + q = 4 + i + j - k,$$

$$pq = (1 + i + j - k)(2 - i - j + k) = 5 + i + j - k,$$

$$p_0 = 1, \quad q_0 = 2, \quad \mathbf{p} = (1, 1, -1), \quad \mathbf{q} = (-1, -1, 1),$$

$$[p, q] = p_0 q_0 - \mathbf{pq} + p_0\mathbf{q}$$
$$+ q_0\mathbf{p} = 2 - (1, 1, -1)(-1, -1, 1) + (-1, -1, 1)$$
$$+ 2(1, 1, -1) = 5 + (1, 1, -1) = 5 + i + j - k$$

$$[p, q] + p = 6 + 2i + 2j - 2k,$$

$$\mathrm{Sc}([p, q] + p) = 6,$$

$$\mathrm{Vec}([p, q] + p) = (2, 2, -2),$$

$$q \cdot ([p, q] + p) = q_0\mathrm{Sc}([p, q] + p) + \mathbf{q}\mathrm{Vec}([p, q] + p)$$
$$= 12 + (-1, -1, 1)(2, 2, -2) = 6,$$

$$(p, q) = p_0 q_0 - q_0\mathbf{p} - \mathbf{p} \times \mathbf{q}$$
$$= (-1, -1, 1) - 2(1, 1, -1) - (1, 1, -1) \times (-1, -1, 1)$$
$$= (-3, -3, 3) = -3i - 3j + 3k.$$

Therefore

$$2p + q - pq + q \cdot ([p, q] + p) + (p, q)$$
$$= 4 + i + j - k - (5 + i + j - k) + 6 - 3i - 3j + 3k$$
$$= 5 - 3i - 3j + 3k.$$

37. We have $p_0 = a_1, q_0 = a_2, \mathbf{p} = (b_1, c_1, d_1), \mathbf{q} = (b_2, c_2, d_2)$, and

$$
\begin{aligned}
[p, q] &= p_0 q_0 - \mathbf{pq} + p_0 \mathbf{q} + q_0 \mathbf{p} \\
&= a_1 a_2 - (b_1, c_1, d_1)(b_2, c_2, d_2) + a_1(b_2, c_2, d_2) + a_2(b_1, c_1, d_1) \\
&= a_1 a_2 - b_1 b_2 - c_1 c_2 - d_1 d_2 + (a_1 b_2, a_1 c_2, a_1 d_2) + \\
&\quad + (a_2 b_1, a_2 c_1, a_1 d_1) \\
&= a_1 a_2 - b_1 b_2 - c_1 c_2 - d_1 d_2 \\
&\quad + (a_1 b_2 + a_2 b_1, a_1 c_2 + a_2 c_1, a_1 d_2 + a_2 d_1), \\
&= a_1 a_2 - b_1 b_2 - c_1 c_2 - d_1 d_2 + (a_1 b_2 + a_2 b_1)i \\
&\quad + (a_1 c_2 + a_2 c_1)j + (a_1 d_2 + a_2 d_1)k.
\end{aligned}
$$

38. Let $p = a_1 + b_1 i + c_1 j + d_1 k$ and $q = a_2 + b_2 i + c_2 j + d_2 k$, where $a_i, b_i, c_i, d_i \in \mathbb{R}, i = 1, 2$. Then

$$
\begin{aligned}
pq &= a_1 a_2 - b_1 b_2 - c_1 c_2 - d_1 d_2 + (a_1 b_2 + b_1 a_2 + c_1 d_2 - d_1 c_2)i + \\
&\quad + (a_1 c_2 - b_1 d_2 + c_1 a_2 + d_1 b_2)j + (a_1 d_2 + b_1 c_2 - c_1 b_2 + d_1 a_2)k, \\
qp &= a_1 a_2 - b_1 b_2 - c_1 c_2 - d_1 d_2 + (a_2 b_1 + b_2 a_1 + c_2 d_1 - d_2 c_1)i + \\
&\quad + (a_2 c_1 - b_2 d_1 + c_2 a_1 + d_2 b_1)j + (a_2 d_1 + b_2 c_1 - c_2 b_1 + d_2 a_1)k,
\end{aligned}
$$

$$
\begin{aligned}
\frac{1}{2}(pq + qp) &= a_1 a_2 - b_1 b_2 - c_1 c_2 - d_1 d_2 + (a_2 b_1 + b_2 a_1)i + \\
&\quad + (a_2 c_1 + a_1 c_2)j + (a_1 d_2 + a_2 d_1)k.
\end{aligned}
$$

Consequently,

$$
[p, q] = \frac{pq + qp}{2}.
$$

39. (a) It is easy to see that

$$
pq = (i + j)(i - j) = -2k, \quad qp = (i - j)(i + j) = 2k,
$$

$$
[p, q] = \frac{pq + qp}{2} = 0_{\mathbb{H}};
$$

(b) $-1 + 2i$;   (c) $3 + i + j + k$.

40. We have

$$
[p, q] = \frac{pq + qp}{2} = \frac{qp + pq}{2} = [q, p].
$$

41. (a) Obviously, it holds that

$$\mathbf{p} = (1, 2, -1), \quad \mathbf{q} = (-1, 1, 1),$$

$$\mathbf{p} \times \mathbf{q} = (3, 0, 3) = 3i + 3k;$$

(b) $3i - 3j$; (c) $-2i - \frac{3}{2}j + \frac{1}{2}k$; (d) k; (e) $-j$; (f) i.

42. We have

$$\mathbf{p} = (b_1, c_1, d_1), \quad \mathbf{q} = (b_2, c_2, d_2),$$

$$\mathbf{p} \times \mathbf{q} = (b_1, c_1, d_1) \times (b_2, c_2, d_2) = (c_1 d_2 - c_2 d_1, b_2 d_1 - b_1 d_2, b_1 c_2 - b_2 c_1),$$

i.e.,

$$p \times q = (c_1 d_2 - c_2 d_1)i + (b_2 d_1 - b_1 d_2)j + (b_1 c_2 - b_2 c_1)k.$$

43. Let $p = a_1 + b_1 i + c_1 j + d_1 k$ and $q = a_2 + b_2 i + c_2 j + d_2 k$, where $a_i, b_i, c_i, d_i \in \mathbb{R}$ $(i = 1, 2)$. Then

$$pq = a_1 a_2 - b_1 b_2 - c_1 c_2 - d_1 d_2 + (a_1 b_2 + b_1 a_2 + c_1 d_2 - c_2 d_1)i +$$
$$+ (a_1 c_2 - b_1 d_2 + c_1 a_2 + d_1 b_2)j + (a_1 d_2 + b_1 c_2 - c_1 b_2 + d_1 a_2)k,$$

$$qp = a_1 a_2 - b_1 b_2 - c_1 c_2 - d_1 d_2 + (a_2 b_1 + b_2 a_1 + c_2 d_1 - d_2 c_1)i +$$
$$+ (a_2 c_1 - b_2 d_1 + c_2 a_1 + d_2 b_1)j + (a_2 d_1 + b_2 c_1 - c_2 b_1 + d_2 a_1)k,$$

Therefore,

$$\frac{1}{2}(pq - qp) = (c_1 d_2 - c_2 d_1)i + (d_1 b_2 - d_2 b_1)j + (b_1 c_2 - c_1 b_2)k,$$

whence

$$p \times q = \frac{pq - qp}{2}.$$

44. (a) We obtain

$$pq = (i + j - k)(1 - i + j + k) = 1 + 3i + j + k,$$

$$qp = (1 - i + j + k)(i + j - k) = 1 - i + j - 3k,$$

$$\frac{1}{2}(pq - qp) = 2i + 2k;$$

(b) $-i - k$, (c) $i - j + k$.

45. (a) We get

$$p_0 = 1, \quad q_0 = 2,$$

$$\mathbf{p} = (-1, 1, 1), \quad \mathbf{q} = (-1, 0, -1),$$

$$p \cdot q = p_0 q_0 + \mathbf{p}\mathbf{q} = 2 + (-1, 1, 1)(-1, 0, -1) = 2,$$

$$\begin{aligned}(p, q) &= p_0 \mathbf{q} - q_0 \mathbf{p} - \mathbf{p} \times \mathbf{q}\\ &= (-1, 0, -1) - 2(-1, 1, 1) - (-1, 1, 1) \times (-1, 0, -1)\\ &= (2, 0, -4) = 2i - 4k,\end{aligned}$$

$$p \times q = \mathbf{p} \times \mathbf{q} = (-1, 1, 1) \times (-1, 0, -1) = (-1, -2, 1),$$

$$p \times q = -i - 2j + k,$$

$$p \cdot q - (p, q) + p \times q = 2 - 2i + 4k - i - 2j + k = 2 - 3i - 2j + 5k;$$

(b) $2 + i + 6j - 5k$; (c) $6 - i - 2j + k$.

46. (a) We have

$$2p - r = 2i + 2j - (i + j + k) = i + j - k,$$

$$p + q = i + j + 1 - i - k = 1 + j - k,$$

$$\mathrm{Sc}(p + q) = 1, \quad \mathrm{Vec}(p + q) = (0, 1, -1),$$

$$r_0 = 0, \quad \mathbf{r} = (1, 1, 1),$$

$$r \cdot (p + q) = r_0 \mathrm{Sc}(p + q) + \mathbf{r}\mathrm{Vec}(p + q) = (1, 1, 1)(0, 1, -1) = 0,$$

$$q + r = 1 + j,$$

$$p(q + r) = (i + j)(1 + j) = -1 + i + j + k,$$

$$2p - r - r \cdot (p + q) - p(q + r) = i + j - k - (-1 + i + j + k) = 1 - 2k;$$

(b) $1 + 2i + 2j + 2k$; (c) $1 + i$; (d) $-2 + 3i - 2k$.

47. (a) We have

$$|p| = \sqrt{9 + 1 + 1 + 1} = 2\sqrt{3},$$

$$\mathrm{sgn}(p) = \frac{p}{|p|} = \frac{\sqrt{3}}{6}(3 + i - j + k) = \frac{\sqrt{3}}{2} + \frac{\sqrt{3}}{6}i - \frac{\sqrt{3}}{6}j + \frac{\sqrt{3}}{6}k;$$

(b) $\frac{\sqrt{3}}{3} - \frac{\sqrt{3}}{3}i + \frac{\sqrt{3}}{3}k$; (c) $\frac{\sqrt{2}}{2}i - \frac{\sqrt{2}}{2}j$.

48. (a) After a straightforward calculation we get

$$pq = (i + j)(1 - k) = 2j,$$

$$|r| = \sqrt{3},$$

$$\operatorname{sgn}(r) = \frac{r}{|r|} = \frac{\sqrt{3}}{3}\, i + \frac{\sqrt{3}}{3}\, j + \frac{\sqrt{3}}{3}\, k,$$

$$pq - p + 2q - 3\operatorname{sgn}(r) = 2 - (1 + \sqrt{3})i + (1 - \sqrt{3})j - (2 + \sqrt{3})k;$$

(b) $(3 + \sqrt{2})i - j;$ (c) $-\sqrt{2}i - \sqrt{2}j - 3\sqrt{2}k;$ (d) $-2 - \frac{\sqrt{3}}{3} - \frac{2\sqrt{3}}{3}i - \frac{\sqrt{3}}{3}k;$
(e) $-2 - \left(1 + \frac{\sqrt{2}}{2}\right)i - \left(3 + \frac{\sqrt{2}}{2}\right)j - 2k;$ (f) $-3i + (\sqrt{2} - 3)j - 3k.$

49. The following calculation solves our exercise:

$$|p| = \sqrt{a^2 + b^2 + c^2 + d^2},$$

$$\operatorname{sgn}(p) = \frac{a}{\sqrt{a^2 + b^2 + c^2 + d^2}} + \frac{b}{\sqrt{a^2 + b^2 + c^2 + d^2}}\, i$$

$$+ \frac{c}{\sqrt{a^2 + b^2 + c^2 + d^2}}\, j + \frac{d}{\sqrt{a^2 + b^2 + c^2 + d^2}}\, k.$$

50. (a) We have

$$p_0 = 1, \quad |p| = \sqrt{6}, \quad \arg(p) = \arccos \frac{\sqrt{6}}{6};$$

(b) $\arccos\left(-\frac{1}{2}\right);$ (c) $\arccos \frac{2}{\sqrt{6}}.$

51. (a) We compute

$$|q| = \sqrt{3}, \quad q_0 = 1, \quad \arg(q) = \arccos \frac{\sqrt{3}}{3},$$

$$p - r = 2, \quad \mathbf{q} = (-1, 0, -1), \quad r_0 = 0, \quad \mathbf{r} = (0, 1, 1),$$

$$[q, r] = q_0 r_0 - \mathbf{q}\mathbf{r} + q_0\mathbf{r} + r_0\mathbf{q} = 1 + (0, 1, 1),$$

$$[q, r] = 1 + j + k,$$

$$\arg(q)(p - r) - [q, r] = 2\arccos \frac{\sqrt{3}}{3} - 1 - j - k;$$

(b) $-\frac{\pi}{2} - i + \frac{\sqrt{2}}{2}j + \left(-4 + \frac{\sqrt{2}}{2}\right)k;$ (c) $3 - \frac{3\pi}{2} + \left(2 + \frac{\pi}{2}\right)i + \left(4 - \frac{\pi}{2}\right)j + 6k.$

52. (a) An easy computation yields

$$|p| = 2, \quad p_0 = 1, \quad \arg(p) = \arccos\frac{1}{2} = \frac{\pi}{3},$$

$$\mathbf{p} = (1,1,1), \quad |\mathbf{p}| = \sqrt{3}, \frac{\mathbf{p}}{|\mathbf{p}|} = \left(\frac{\sqrt{3}}{3}, \frac{\sqrt{3}}{3}, \frac{\sqrt{3}}{3}\right).$$

Consequently,

$$p = 2\left[\cos\frac{\pi}{3} + \left(\frac{\sqrt{3}}{3}i + \frac{\sqrt{3}}{3}j + \frac{\sqrt{3}}{3}k\right)\sin\frac{\pi}{3}\right];$$

(b) $\sqrt{2}\left(\cos\frac{\pi}{2} + \left(\frac{\sqrt{2}}{2}i + \frac{\sqrt{2}}{2}j\right)\sin\frac{\pi}{2}\right);$ (c) $\sqrt{2}\left(\cos\frac{\pi}{2} + \left(\frac{\sqrt{2}}{2}i + \frac{\sqrt{2}}{2}k\right)\sin\frac{\pi}{2}\right);$
(d) $\sqrt{2}\left(\cos\frac{\pi}{2} + \left(\frac{\sqrt{2}}{2}j + \frac{\sqrt{2}}{2}k\right)\sin\frac{\pi}{2}\right);$ (e) $\sqrt{2}\left(\cos\frac{\pi}{2} + \left(\frac{\sqrt{2}}{2}i - \frac{\sqrt{2}}{2}j\right)\sin\frac{\pi}{2}\right);$
(f) $\sqrt{2}\left(\cos\frac{\pi}{2} + \left(\frac{\sqrt{2}}{2}i - \frac{\sqrt{2}}{2}k\right)\sin\frac{\pi}{2}\right);$ (g) $\sqrt{2}\left(\cos\frac{\pi}{2} + \left(\frac{\sqrt{2}}{2}j - \frac{\sqrt{2}}{2}k\right)\sin\frac{\pi}{2}\right).$

53. (a) $p = 1 + i1 + 1j + i1j;$ (b) $p = 2 + i\left(-\frac{3}{2}\right) + (-4)j + i5j;$ (c) $p = \frac{1}{2} + i(-2) + 3j + i(-4)j.$

54. (a) We have

$$p = 1j + i1j,$$

$$q = 1 + i(-1) + i(-1)j,$$

$$pq = (1j + i1j)(1 + i(-1) + i(-1)j) = 1 + i(-1) + i2j;$$

(b) $\frac{3}{2} + i\left(-\frac{1}{2}\right) + \left(-\frac{1}{2}\right)j + i\left(-\frac{11}{2}\right)j;$ (c) $-1 + i(6) + (-5)j + i5j.$

55. (a) $1 + i2 + 4j + i3j;$ (b) $1 + i1 + 1j + i1j;$ (c) $1 + i2 + 4j + i1j.$

56. (a) It is clear that

$$ipj = i\left(2 + \frac{3}{2}i - \frac{1}{2}j + k\right)j = 1 + \frac{1}{2}i - \frac{3}{2}j + 2k,$$

$$p + ipj = 3 + 2i - 2j + 3k,$$

$$p - ipj = 1 + i + j - k,$$

$$p_+ = \frac{3}{2} + i - j + \frac{3}{2}k, \quad p_- = \frac{1}{2} + \frac{1}{2}i + \frac{1}{2}j - \frac{1}{2}k;$$

(b) $p_+ = \frac{3}{2} + i - j + \frac{3}{2}k,$ $p_- = -\frac{1}{2} + \frac{1}{2}k;$ (c) $p_+ = -\frac{1}{2} - i + j - \frac{1}{2}k,$
$p_- = \frac{3}{2} - \frac{3}{2}k.$

57. We have

$$ipj = d - ci - bj + ak,$$

$$p + ipj = a + d + (b - c)i - (b - c)j + (a + d)k,$$

$$p - ipj = a - d + (b + c)i + (b + c)j - (a - d)k,$$

$$p_+ = \frac{a+d}{2} + \frac{b-c}{2}i - \frac{b-c}{2}j + \frac{a+d}{2}k,$$

$$p_- = \frac{a-d}{2} + \frac{b+c}{2}i + \frac{b+c}{2}j - \frac{a-d}{2}k.$$

58. We have

$$
\begin{aligned}
(a + d + i(b - c))\frac{1+k}{2} &= \frac{a + ak + d + dk + i(b - c) - j(b - c)}{2} \\
&= \frac{a+d}{2} + \frac{b-c}{2}i - \frac{b-c}{2}j + \frac{a+d}{2}k \\
&= p_+,
\end{aligned}
$$

where in the last equality we used the previous exercise. Also,

$$
\begin{aligned}
(a - d + i(b + c))\frac{1-k}{2} &= \frac{a - d - (a - d)k + i(b + c) + j(b + c)}{2} \\
&= \frac{a-d}{2} + \frac{b+c}{2}i + \frac{b+c}{2}j - \frac{a-d}{2}k \\
&= p_-.
\end{aligned}
$$

59. We have

$$
\begin{aligned}
\frac{1+k}{2}(a + d + j(c - b)) &= \frac{a + d + j(c - b) + (a + d)k - i(c - b)}{2} \\
&= \frac{a+d}{2} + \frac{b-c}{2}i - \frac{b-c}{2}j + \frac{a+d}{2}k \\
&= p_+,
\end{aligned}
$$

$$
\begin{aligned}
\frac{1-k}{2}(a - d + j(b + c)) &= \frac{(a - d) + j(b + c) - k(a - d) + i(b + c)}{2} \\
&= p_-.
\end{aligned}
$$

60. Let $p = a + bi + cj + dk$ $(a, b, c, d \in \mathbb{R})$. Then

$$|p_+|^2 = \frac{a^2 + b^2 + c^2 + d^2 + 2ad - 2bc}{2},$$

$$|p_-|^2 = \frac{a^2 + b^2 + c^2 + d^2 - 2ad + 2bc}{2}.$$

Consequently

$$|p_+|^2 + |p_-|^2 = a^2 + b^2 + c^2 + d^2 = |p|^2.$$

62. (a) $0_{\mathbb{H}}$;   (b) $1 - i - j - k$;   (c) 4;   (d) $j + k$;   (e) $2 - 2k$.

64. Note that

$$\mathrm{Sc}\,(p\mathbf{q}\overline{p}) = \frac{p\mathbf{q}\overline{p} + \overline{p\mathbf{q}\overline{p}}}{2} = \frac{p}{2}\,(\mathbf{q} + \overline{\mathbf{q}})\,\overline{p} = 0.$$

66. The following computation leads to the solution:

$$|q\mathbf{p}\overline{q}| = |q|\,|\mathbf{p}|\,|\overline{q}| = |\mathbf{p}|\,|\overline{q}| = |\mathbf{p}|\,.$$

68. Indeed,

$$q\,(\alpha\mathbf{p} + \mathbf{q})\,\overline{q} = q\alpha\mathbf{p}\overline{q} + q\mathbf{q}\overline{q} = \alpha q\mathbf{p}\overline{q} + q\mathbf{q}\overline{q}.$$

70. (a) $-\frac{2}{\sqrt{3}}i + \frac{2}{\sqrt{3}}j - \frac{2}{\sqrt{3}}k$;   (b) i;   (c) j;   (d) k;   (e) $101\left(-\frac{i}{6} + \frac{2j}{\sqrt{6}} - \frac{k}{\sqrt{6}}\right)$;
(f) $2\left(-\frac{i}{\sqrt{19}} + \frac{3j}{\sqrt{19}} - \frac{3k}{\sqrt{19}}\right)$.

Chapter 2

1. (a) -3;   (b) -7;   (c) 343.
2. (a) $\sqrt{7}$;   (b) $\sqrt{15}$;   (c) 7.
3. $2 + 2i + 2j - 2k$.
4. $p = \frac{1}{4}i$.
5. (b) $R(\mathbf{n}, \theta)$

$$= \begin{pmatrix} (1 - \cos\theta)n_1^2 + \cos\theta & (1 - \cos\theta)n_1 n_2 - \sin\theta n_3 & (1 - \cos\theta)n_1 n_3 + \sin\theta n_2 \\ (1 - \cos\theta)n_1 n_2 + \sin\theta n_3 & (1 - \cos\theta)n_2^2 + \cos\theta & (1 - \cos\theta)n_2 n_3 - \sin\theta n_1 \\ (1 - \cos\theta)n_1 n_3 + \sin\theta n_1 & (1 - \cos\theta)n_2 n_3 - \sin\theta n_2 & (1 - \cos\theta)n_3^2 + \cos\theta \end{pmatrix};$$

(c)

$$N = \begin{pmatrix} 0 & -n_3 & n_2 \\ n_3 & 0 & -n_1 \\ -n_2 & n_1 & 0 \end{pmatrix};$$

(d)

$$N^2 = \begin{pmatrix} n_1^2 - 1 & n_1 n_2 & n_1 n_3 \\ n_1 n_2 & n_2^2 - 1 & n_2 n_3 \\ n_1 n_3 & n_2 n_3 & n_3^2 - 1 \end{pmatrix} = \mathbf{nn} - I;$$

(e) **Hint.** Use the definition of the matrix N and (c);
(f) **Hint.** Use (b), (c), (d), (e);
(g) **Hint.** Use the definition of $R(\mathbf{n}, \cdot)$;
(h) From (f) we get

$$R(\mathbf{n}, \epsilon) = I + \epsilon N + O(\epsilon^2).$$

Then

$$R(\mathbf{n}, \theta) = \left(I + \frac{\theta}{k} N + O\left(\frac{1}{k^2}\right) \right)^k \to e^{\theta N} \text{ as } k \to \infty;$$

(i)

$$e^{\theta N} = I + \theta N + \frac{\theta^2}{2} N^2 + \frac{\theta^3}{3!} N^3 + \frac{\theta^4}{4!} N^4 + \frac{\theta^5}{5!} N^5 + \cdots$$

$$= I + \theta N + \frac{\theta^2}{2} N^2 - \frac{\theta^3}{3!} N + \frac{\theta^4}{4!} N^2 + \frac{\theta^5}{5!} N - \cdots$$

$$= I + \sin\theta N + (1 - \cos\theta) N^2$$

$$= I + \sin\theta N + (1 - \cos\theta)(\mathbf{nn} - I).$$

7. Note that

$$\bar{q} = \cos\frac{\theta}{2} - \sin\frac{\theta}{2}\mathbf{u},$$

$$q\mathbf{v}\bar{q} = \left(\cos\frac{\theta}{2} + \sin\frac{\theta}{2}\mathbf{u} \right) \mathbf{v} \left(\cos\frac{\theta}{2} - \sin\frac{\theta}{2}\mathbf{u} \right)$$

$$= \cos^2\frac{\theta}{2}\mathbf{v} + \sin\frac{\theta}{2}\cos\frac{\theta}{2}(\mathbf{uv} - \mathbf{vu}) - \sin^2\frac{\theta}{2}\left(\mathbf{v}(\mathbf{u}\cdot\mathbf{u}) - 2\mathbf{u}(\mathbf{u}\cdot\mathbf{v}) \right),$$

from this, the previous exercise and the fact that $|\mathbf{u}| = 1$ we obtain

$$q\mathbf{v}\overline{q} = \left(\cos^2\frac{\theta}{2} - \sin^2\frac{\theta}{2}\right)\mathbf{v} + \sin\theta(\mathbf{u}\times\mathbf{v}) + 2\sin^2\frac{\theta}{2}\mathbf{u}(\mathbf{u}\cdot\mathbf{v})$$

$$= \cos\theta\mathbf{v} + \sin\theta(\mathbf{u}\times\mathbf{v}) + (1 - \cos\theta)\mathbf{u}(\mathbf{u}\cdot\mathbf{v}).$$

The last equality yields $\mathbf{v}' = q\mathbf{v}\overline{q}$.

8. (a) j;   (b) k;   (c) i;   (d) $j+k$;   (e) $i+j$;   (f) $i+k$;   (g) $i+j+k$;   (h) $-i+j-k$;   (i) $i+2j-3k$;   (j) $-1-4i+j+k$;   (k) $ci+aj+bk$;   (l) $a_0+ci+aj+bk$.

9. (a) $\frac{1}{4}i + \frac{1}{2}j + \left(-\frac{1}{4} - \frac{\sqrt{2}}{2}\right)k$;   (b) $\left(-\frac{1}{2} - \frac{\sqrt{2}}{4}\right)i + \frac{1}{2}j + \left(-\frac{1}{2} + \frac{\sqrt{2}}{4}\right)k$;

(c) $\left(\frac{\sqrt{2}}{2} - \frac{1}{4}\right)i + \left(\frac{1}{2} + \frac{\sqrt{2}}{4}\right)j + \frac{1}{4}k$;   (d) $\left(\frac{1}{2} + \frac{3\sqrt{2}}{4}\right)i + \frac{1}{2}j - \frac{1}{2}k$;   (e) $1 +$

$\left(\frac{\sqrt{2}}{4} - \frac{1}{2}\right)i + \frac{3}{2}j - \frac{1}{2}k$,

(f) $i\left(\frac{a}{4} - \frac{b}{2} - \frac{b\sqrt{2}}{4} + \frac{c\sqrt{2}}{2} - \frac{c}{4}\right) + j\left(\frac{a}{2} - \frac{a\sqrt{2}}{4} + \frac{b}{2} + \frac{c}{2} + \frac{c\sqrt{2}}{4}\right)$

$\quad + k\left(-\frac{a\sqrt{2}}{2} - \frac{a}{4} - \frac{b}{2} + \frac{b\sqrt{2}}{4} + \frac{c}{4}\right)$;

(g) $\quad a_0 + i\left(\frac{a}{4} - \frac{b}{2} - \frac{b\sqrt{2}}{4} + \frac{c\sqrt{2}}{2} - \frac{c}{4}\right)$

$\quad + j\left(\frac{a}{2} - \frac{a\sqrt{2}}{4} + \frac{b}{2} + \frac{c}{2} + \frac{c\sqrt{2}}{4}\right) + k\left(-\frac{a\sqrt{2}}{2} - \frac{a}{4} - \frac{b}{2} + \frac{b\sqrt{2}}{4} + \frac{c}{4}\right)$.

10. (a) $-12 - 70i + 24j - 26k$;   (b) $-147 + 91i - 9j + 29k$.

11. We have:

(a) $(a_1d_2 - a_2d_1 - b_1c_2 + b_2c_1)i + (a_1b_2 - a_2b_1 - c_1d_2 + c_2d_1)j$
$\quad + (a_1c_2 - a_2c_1 - b_2d_1 + b_1d_2)k$;

(b) $a_1a_2 - b_1b_2 - c_1c_2 - d_1d_2 + (a_1d_2 + a_2d_1)i$
$\quad + (a_1b_2 + a_2b_1)j + (a_1c_2 + a_2c_1)k$;

(c) $(b_1c_2 - b_2c_1)i + (c_1d_2 - c_2d_1)j + (b_2d_1 - b_1d_2)k$;

(d) $\dfrac{a_1}{\sqrt{a_1^2+b_1^2+c_1^2+d_1^2}} + \dfrac{d_1}{\sqrt{a_1^2+b_1^2+c_1^2+d_1^2}}i + \dfrac{b_1}{\sqrt{a_1^2+b_1^2+c_1^2+d_1^2}}j + \dfrac{c_1}{\sqrt{a_1^2+b_1^2+c_1^2+d_1^2}}k$;

(e) $\frac{a_1+d_1}{2} + \frac{a_1+d_1}{2}i + \frac{b_1-c_1}{2}j - \frac{b_1-c_1}{2}k$;

(f) $\frac{a_1-d_1}{2} - \frac{a_1-d_1}{2}i + \frac{b_1+c_1}{2}j + \frac{b_1+c_1}{2}k$.

12. $-2 - 2i + 3j$.

13. (a) $1 - 2i + 2j + 3k$;   (b) $\cosh(1) + \tanh(1)i + \sinh(1)j$.

15. The multiplication of unit quaternions corresponds to the composition of three-dimensional rotations. Therefore this property can be made intuitive by showing that three-dimensional rotations do not commute in general.

16. The matrix is given by

$$R(q) = \begin{pmatrix} q_0^2 + q_1^2 - q_2^2 - q_3^2 & 2(-q_0q_3 + q_1q_2) & 2(q_0q_2 + q_1q_3) \\ 2(q_0q_3 + q_2q_1) & q_0^2 - q_1^2 + q_2^2 - q_3^2 & 2(-q_0q_1 + q_2q_3) \\ 2(-q_0q_2 + q_3q_1) & 2(q_0q_1 + q_2q_3) & q_0^2 - q_1^2 - q_2^2 + q_3^2 \end{pmatrix}.$$

17. Let us consider the 4D space with a basis $(1, i, j, k)$. Then two vectors $p_1 = a_1 + b_1 i + c_1 j + d_1 k$, $p_2 = a_2 + b_2 i + c_2 j + d_2 k$, $a_i, b_i, c_i, d_i \in \mathbb{R}$ $(i = 1, 2)$ are called perpendicular if $p_1 \cdot p_2 = 0$. Let the given triangle be $\triangle ABC$ and $\angle ACB = 90°$. Let also

$$p_1 = \overrightarrow{BC} = a_1 + b_1 i + c_1 j + d_1 k,$$

$$p_2 = \overrightarrow{CA} = a_2 + b_2 i + c_2 j + d_2 k.$$

Then

$$p_3 = \overrightarrow{BA} = \overrightarrow{BC} + \overrightarrow{CA} = p_1 + p_2.$$

Dot squaring we have

$$p_3 \cdot p_3 = (p_1 + p_2) \cdot (p_1 + p_2) = p_1 \cdot p_1 + 2 p_1 \cdot p_2 + p_2 \cdot p_2.$$

Since $\angle BCA = 90°$, $p_1 \cdot p_2 = 0$. Therefore,

$$p_3 \cdot p_3 = p_1 \cdot p_1 + p_2 \cdot p_2$$

or $|p_3|^2 = |p_1|^2 + |p_2|^2$, i.e.,

$$AB^2 = BC^2 + CA^2.$$

18. Here we will use the notations in the solution of the previous exercise. Let CM, $M \in AB$ is the median to the hypothenuse AB. Then

$$p_4 = \overrightarrow{CM} = \frac{1}{2}(-p_1 + p_2).$$

Squaring dot

$$p_4 \cdot p_4 = \frac{1}{4}(-p_1 + p_2) \cdot (-p_1 + p_2),$$

$$|p_4|^2 = \frac{1}{4}(|p_1|^2 - 2 p_1 \cdot p_2 + |p_2|^2) = \frac{1}{4}(|p_1|^2 + |p_2|^2) = \frac{1}{4}|p_3|^2,$$

i.e.

$$|p_4| = \frac{1}{2}|p_3|.$$

Consequently,

$$CM = \frac{1}{2}AB.$$

Chapter 3

1. This is not always correct. Take, for instance, $p_n = (-1)^n$.

4. (a) $0_{\mathbb{H}}$;   (b) $0_{\mathbb{H}}$;   (c) $0_{\mathbb{H}}$;

$$
\text{(d)} \quad \lim_{n \longrightarrow \infty} \frac{n + i + j + k}{n - i + k} = \lim_{n \longrightarrow \infty} \frac{(n + i + j + k)(n + i - k)}{(n - i + k)(n + i - k)}
$$

$$
= \lim_{n \longrightarrow \infty} \frac{n^2 + (2n - 1)i + (n + 2)j - k}{n^2 + 2}
$$

$$
= \lim_{n \longrightarrow \infty} \frac{n^2}{n^2 + 2} + \lim_{n \longrightarrow \infty} \frac{2n - 1}{n^2 + 1} i
$$

$$
+ \lim_{n \longrightarrow \infty} \frac{n + 2}{n^2 + 2} j - \lim_{n \longrightarrow \infty} \frac{1}{n^2 + 2} k
$$

$$
= 1_{\mathbb{H}};
$$

(e) $2i$;   (f) k;   (g) $1_{\mathbb{H}}$;   (h) ∞;   (i) ∞;

$$
\text{(j)} \quad \lim_{n \longrightarrow \infty} \frac{1 + 2n^2 i - j - k}{n^2 + j + k} = \lim_{n \longrightarrow \infty} \frac{(1 + 2n^2 i - j - k)(n^2 - j - k)}{(n^2 + j + k)(n^2 - j - k)}
$$

$$
= \lim_{n \longrightarrow \infty} \frac{n^2 - 2}{n^4 + 2} + \frac{2n^4}{n^4 + 2} i + \frac{n^2 - 1}{n^4 + 2} j
$$

$$
+ \frac{-3n^2 - 1}{n^4 + 2} k
$$

$$
= 2i;
$$

(k) k;   (l) $1_{\mathbb{H}}$;   (m) ∞.

5. $0_{\mathbb{H}}$.

6. $\frac{1}{6} + \frac{1}{6}i - \frac{1}{6}k$.

7. $\frac{1}{6} - \frac{1}{12}i - \frac{1}{9}j + \frac{1}{3}k$.

8. (a)–(g) $1_{\mathbb{H}}$.

9. $1 + i + e^{\frac{1}{3}}k$.

10. $1 + i + \frac{1}{l+1}k$.

11. $1 + \frac{1}{2}i + 2j + 2k$.

12. (a) $1_{\mathbb{H}}$;   (b) $1_{\mathbb{H}}$;   (c) $1_{\mathbb{H}}$;   (d) $\frac{-4 - 2i + 2j + 2k}{7}$;   (e) $\frac{11 + 2i + 5j + 3k}{11}$;
    (f) $\frac{2 - 2i + k}{3}$;   (g) $\frac{1570 + 250i + 6j + 275k}{7817}$;   (h) $\frac{9 - i - j - 6k}{13}$;   (i) $\frac{-8i + 5j - k}{6}$;
    (j) $\frac{-2 + 2i - k}{9}$.

15. (a) $1 + i + j$;   (b) $\frac{1}{4}j + \frac{1}{9}k$.

16. (a) $\frac{1}{2}j + \frac{1}{2}k$;   (b) $-\frac{1}{2}j - \frac{1}{2}k$;   (c) $0_{\mathbb{H}}$;   (d) no solutions;   (e) $-\frac{1}{2} - \frac{1}{2}k$;
    (f) $i + 1$;   (g) $\frac{1}{2} + \frac{1}{2}k$.

Chapter 4

2. For the necessity condition, assume that $\sum_{n=1}^{\infty} p_n$ is convergent. Then the sequence $S_N = \sum_{n=1}^{N} p_n$ is convergent. From here it follows that the sequences

$$\sum_{n=1}^{N} a_n, \quad \sum_{n=1}^{N} b_n, \quad \sum_{n=1}^{N} c_n, \quad \text{and} \quad \sum_{n=1}^{N} d_n$$

converge. Therefore

$$\sum_{n=1}^{\infty} a_n, \quad \sum_{n=1}^{\infty} b_n, \quad \sum_{n=1}^{\infty} c_n, \quad \sum_{n=1}^{\infty} d_n$$

are convergent. Conversely, let

$$\sum_{n=1}^{\infty} a_n, \quad \sum_{n=1}^{\infty} b_n, \quad \sum_{n=1}^{\infty} c_n, \quad \sum_{n=1}^{\infty} d_n$$

be convergent. Then the sequences

$$\sum_{n=1}^{N} a_n, \quad \sum_{n=1}^{N} b_n, \quad \sum_{n=1}^{N} c_n, \quad \sum_{n=1}^{N} d_n$$

are convergent, and moreover

$$\sum_{n=1}^{N} (a_n + b_n i + c_n j + d_n k)$$

is convergent. Hence, $\sum_{n=1}^{\infty} p_n$ is convergent.

4. (a) $\frac{23}{90} + \frac{1}{4}i + \frac{\pi}{4}j + \frac{1}{2}k$; (b) $1 + i - j + k$.

5. $(1_{\mathbb{H}} + p + p^2 + \cdots)^m$.

6. We have
   (a) Absolutely convergent for $\alpha \in \left(\frac{1}{2}, 1\right)$, divergent for $\alpha \leq \frac{1}{2}, \alpha \geq 1$;
   (b) Absolutely divergent for $\alpha > 5$, divergent for $\alpha \leq 5$;
   (c) Absolutely convergent for $\alpha < -1$, divergent for $\alpha \geq -1$;
   (d) Absolutely convergent for for $\alpha > \frac{1}{2}$, divergent for $\alpha \leq \frac{1}{2}$;
   (e) Absolutely divergent for for $\alpha < -1$, divergent for $\alpha \geq -1$;
   (f) Absolutely convergent for $\alpha + \beta > \frac{1}{2}$, divergent for $\alpha + \beta \leq \frac{1}{2}$.

7. (a) yes; (b) yes; (c) yes; (d) no; (e) no; (f) no.

8. (a) convergent; (b) convergent; (c) divergent; (d) convergent; (e) convergent; (f) divergent; (g) convergent; (h) divergent; (i) convergent for

$\alpha > \frac{1}{2}$ and divergent for $0 < \alpha < \frac{1}{2}$;   (j) convergent;   (k) convergent for $p > \frac{1}{2}$ and divergent for $0 < p < \frac{1}{2}$;   (l) divergent;   (m) convergent; (n) divergent;   (o) divergent.

Chapter 5

1. (a) We have

$$p_0 = 1, \quad \mathbf{p} = -i + j = (-1, 1, 0),$$

$$|\mathbf{p}| = \sqrt{2}, \quad \mathrm{sgn}(\mathbf{p}) = \frac{-i + j}{\sqrt{2}} = -\frac{\sqrt{2}}{2}i + \frac{\sqrt{2}}{2}j,$$

$$e^{1-i+j} = e\cos\sqrt{2} - e\frac{\sqrt{2}}{2}\sin\sqrt{2}i + e\frac{\sqrt{2}}{2}\sin\sqrt{2}j.$$

Further, we obtain

$$\text{(b) } e^2\cos\sqrt{2} + e^2\frac{\sqrt{2}}{2}\sin\sqrt{2}i - e^2\frac{\sqrt{2}}{2}\sin\sqrt{2}j;$$

$$\text{(c) } \cos\sqrt{2} + \frac{\sqrt{2}}{2}\sin\sqrt{2}i - \frac{\sqrt{2}}{2}\sin\sqrt{2}k;$$

(d) $\cos 1 + i\sin 1$;   (e) $\cos 1 + \sin 1 j$;   (f) $\cos 1 + \sin 1 k$.

2. (1) For the solutions the following relations are important:

$$e^{-i} = \cos(1) - i\sin(1), \quad e^{j} = \cos(1) + j\sin(1),$$

$$e^{-i}\frac{1+k}{2} = \frac{\cos(1) - \sin(1)i + \sin(1)j + \cos(1)k}{2},$$

$$\frac{1+k}{2}e^{j} = \frac{\cos(1) - \sin(1)i + j\sin(1) + \cos(1)k}{2};$$

(2)–(40) **Hint.** Use the solution of (1).

6. (a) First, direct computations show that

$$pq = (1 + i + j)(i + k) = -1 + 2i - j, \quad \bar{r} = -j - k, \quad q_0 = 0,$$

$$\mathbf{q} = j + k, \quad |\mathbf{q}| = \sqrt{2}, \quad \mathrm{sgn}(\mathbf{q}) = \tfrac{\sqrt{2}}{2}j + \tfrac{\sqrt{2}}{2}k,$$

$$e^{r} = \cos\sqrt{2} + \tfrac{\sqrt{2}}{2}\sin\sqrt{2}j + \tfrac{\sqrt{2}}{2}\sin\sqrt{2}k,$$

$$[p, q] = p_0 q_0 - \mathbf{pq} + p_0\mathbf{q} + q_0\mathbf{p} = -1 + i + k.$$

Then

$$pq - \bar{r} + e^q + [p, q] = -2 + \cos\sqrt{2} + 3i + \frac{\sqrt{2}}{2}\sin\sqrt{2}\,j$$

$$+ \left(2 + \frac{\sqrt{2}}{2}\cos\sqrt{2}\right)k.$$

(b) $2 + \cos\sqrt{2} + 2i + \left(-1 + \frac{\sqrt{2}}{2}\sin\sqrt{2}\right)j + \left(3 + \frac{\sqrt{2}}{2}\sin\sqrt{2}\right)k;$

(c) $\left(1 + \dfrac{\pi}{2} + e\cos\sqrt{2}\right) + \left(\dfrac{\pi}{2} - \dfrac{\sqrt{2}}{2}e\sin\sqrt{2}\right)i$

$\qquad + \left(1 + \dfrac{\pi}{2} - \dfrac{\sqrt{2}}{2}e\cos\sqrt{2}\right)j - k.$

7. (a) We have

$$|p| = 2, \quad \mathbf{p} = (1, 1, 1) = i + j + k, \quad |\mathbf{p}| = \sqrt{3},$$

$$\operatorname{sgn}(\mathbf{p}) = \tfrac{\sqrt{3}}{3}i + \tfrac{\sqrt{3}}{3}j + \tfrac{\sqrt{3}}{3}k,$$

$$\arg(p) = \arccos\tfrac{1}{2} = \tfrac{\pi}{3}, \quad \ln(p) = \ln 2 + \tfrac{\sqrt{3}\pi}{9}i + \tfrac{\sqrt{3}\pi}{9}j + \tfrac{\sqrt{3}\pi}{9}k.$$

For the remaining cases the results are

(b) $\dfrac{\pi}{2}i;$ (c) $\dfrac{\pi}{2}j;$

(d) $k\dfrac{\pi}{2};$ (e) $\ln\sqrt{2} + \dfrac{\pi\sqrt{2}}{4}i + \dfrac{\pi\sqrt{2}}{4}j;$

(f) $\ln\sqrt{2} + \dfrac{\pi\sqrt{2}}{4}i + \dfrac{\pi\sqrt{2}}{4}k;$ (g) $\ln\sqrt{2} + \dfrac{\pi\sqrt{2}}{4}j + \dfrac{\pi\sqrt{2}}{4}k.$

8. (a) It is easy to prove that

$$|p| = 2, \quad \mathbf{p} = (-1, -1, 1) = -i - j + k, \quad \operatorname{sgn}(\mathbf{p}) = -\tfrac{\sqrt{3}}{3}i - \tfrac{\sqrt{3}}{3}j + \tfrac{\sqrt{3}}{3}k,$$

$$\arg(p) = \arccos\tfrac{1}{2} = \tfrac{\pi}{3}, \quad \ln(p) = \ln 2 - \tfrac{\sqrt{3}\pi}{9}i - \tfrac{\sqrt{3}\pi}{9}j + \tfrac{\sqrt{3}\pi}{9}k.$$

Then the first matrix representation is

$$\begin{pmatrix} \ln 2 - \frac{\sqrt{3}\pi}{9}i & \frac{\sqrt{3}\pi}{9} - \frac{\sqrt{3}\pi}{9}i \\ -\frac{\sqrt{3}\pi}{9} - \frac{\sqrt{3}\pi}{9}i & \ln 2 + \frac{\sqrt{3}\pi}{9}i \end{pmatrix}.$$

The second matrix representation of $\ln(p)$ is

$$\begin{pmatrix} \ln 2 & \frac{\sqrt{3}\pi}{9} & \frac{\sqrt{3}\pi}{9} & \frac{\sqrt{3}\pi}{9} \\ -\frac{\sqrt{3}\pi}{9} & \ln 2 & \frac{\sqrt{3}\pi}{9} & -\frac{\sqrt{3}\pi}{9} \\ -\frac{\sqrt{3}\pi}{9} & -\frac{\sqrt{3}\pi}{9} & \ln 2 & \frac{\sqrt{3}\pi}{9} \\ -\frac{\sqrt{3}\pi}{9} & \frac{\sqrt{3}\pi}{9} & -\frac{\sqrt{3}\pi}{9} & \ln 2 \end{pmatrix}.$$

The solutions for the other problems are

(b)
$$\begin{pmatrix} \ln 4 & \frac{\sqrt{3}\pi}{6} + \frac{\pi}{6}i \\ -\frac{\sqrt{3}\pi}{6} + \frac{\pi}{6}i & \ln 4 \end{pmatrix}, \qquad \begin{pmatrix} \ln 4 & \frac{\sqrt{3}\pi}{6} & 0 & -\frac{\pi}{6} \\ -\frac{\sqrt{3}\pi}{6} & \ln 4 & -\frac{\pi}{6} & 0 \\ 0 & \frac{\pi}{6} & \ln 4 & \frac{\sqrt{3}\pi}{6} \\ \frac{\pi}{6} & 0 & -\frac{\sqrt{3}\pi}{6} & \ln 4 \end{pmatrix};$$

(c)
$$\begin{pmatrix} \ln\sqrt{2} - \frac{\pi}{4}i & 0 \\ 0 & \ln\sqrt{2} + \frac{\pi}{4}i \end{pmatrix}, \qquad \begin{pmatrix} \ln\sqrt{2} & 0 & \frac{\pi}{4} & 0 \\ 0 & \ln\sqrt{2} & 0 & -\frac{\pi}{4} \\ -\frac{\pi}{4} & 0 & \ln\sqrt{2} & 0 \\ 0 & \frac{\pi}{4} & 0 & \ln\sqrt{2} \end{pmatrix}.$$

9. (a) We present the calculations step by step:

$$\operatorname{sgn}(\mathbf{p}) = -\frac{\sqrt{3}}{3}i + \frac{\sqrt{3}}{3}j - \frac{\sqrt{3}}{3}k, \qquad p_0 = 1, \quad |p| = 2,$$

$$\arg(p) = \arccos\frac{1}{2} = \frac{\pi}{3}, \qquad \ln(p) = -\frac{\pi\sqrt{3}}{9}i + \frac{\pi\sqrt{3}}{9}j - \frac{\pi\sqrt{3}}{9}k + \ln 2,$$

$$q_0 = 0, \quad \arg(q) = \arccos 0 = \frac{\pi}{2}, \qquad r_0 = 0, \quad \arg(r) = \arccos 0 = \frac{\pi}{2},$$

$$\operatorname{sgn}(\mathbf{r}) = \frac{\sqrt{2}}{2}i + \frac{\sqrt{2}}{2}k, \qquad \ln(r) = \ln\sqrt{2} + \frac{\pi\sqrt{2}}{4}i + \frac{\pi\sqrt{2}}{4}k,$$

$$\arg(q)\ln(r) = \frac{\pi}{2}\ln\sqrt{2} + \frac{\pi^2\sqrt{2}}{8}i + \frac{\pi^2\sqrt{2}}{8}k, \qquad \mathbf{q} = i + j = (1,1,0),$$

and

$$[p,q] = p_0 q_0 - \mathbf{p}\mathbf{q} + p_0\mathbf{q} + q_0\mathbf{p} = (1,1,0),$$

$$[p,q] = i + j, \qquad p \cdot r = p_0 r_0 + \mathbf{p}\mathbf{r} = -2.$$

Then

$$\ln(p) + \arg(q)\ln(r) + [p,q] - p \cdot r$$

$$= \left(-2 + \ln 2 + \frac{\pi}{2}\ln\sqrt{2}\right) + \left(1 - \frac{\pi\sqrt{3}}{9} + \frac{\pi^2\sqrt{2}}{8}\right)i$$

$$+ \left(1 + \frac{\pi\sqrt{3}}{9}\right)j + \left(\frac{\pi^2\sqrt{2}}{8} - \frac{\pi\sqrt{3}}{9}\right)k.$$

The remaining exercises have the following solutions:

(b) $\quad \ln\sqrt{2} + \frac{\pi\sqrt{2}}{4}i + \left(\frac{\pi\sqrt{2}}{4}\right)j + 2k,$

(c) $\quad \ln 2 - \ln\sqrt{2} + 2 + \left(-\frac{\pi\sqrt{3}}{9} - \frac{\pi\sqrt{2}}{4} + 2\right)i$

$$+ \left(1 + \frac{\pi\sqrt{3}}{9} - \frac{\pi\sqrt{2}}{4}\right)j + \left(-1 - \frac{\pi\sqrt{3}}{9}\right)k.$$

14. (a) One gets

$$(1+i)^2 = e^{\ln(1+i)2} = e^{2\ln\sqrt{2}+i\frac{\pi}{2}} = e^{\ln 2 + i\frac{\pi}{2}} = 2i;$$

(b) $-8 + 8i + 8j - 8k;$ (c) $-2^9 - 2^9 i - 2^9 j - 2^9 k.$

15. (a) It is easy to verify that

$$2^{1-i+j+k} = e^{\ln 2(1-i+j+k)}$$

$$= e^{\ln 2 - \ln 2i + \ln 2j + \ln 2k}$$

$$= 2\left(\cos(\sqrt{3}\ln 2) - \frac{\sqrt{3}}{3}\sin(\sqrt{3}\ln 2)i + \frac{\sqrt{3}}{3}\sin(\sqrt{3}\ln 2)j\right.$$

$$\left. + \frac{\sqrt{3}}{3}\sin(\sqrt{3}\ln 2)k\right).$$

In an analogous way one computes

(b) $3\cos(\ln 3) - 3\sin(\ln 3)i;$

(c) $\cos(\sqrt{3}\ln 4) + \frac{\sqrt{3}}{3}\sin(\sqrt{3}\ln 4)i + \frac{\sqrt{3}}{3}\sin(\sqrt{3}\ln 4)j$

$$+ \frac{\sqrt{3}}{3}\sin(\sqrt{3}\ln 4)k;$$

(d) $e \cos 1 + e \sin 1 i$;   (e) $e \cos \sqrt{2} + e \dfrac{\sqrt{2}}{2} \sin \sqrt{2} i + e \dfrac{\sqrt{2}}{2} \sin \sqrt{2} j$;

(f) $e \cos \sqrt{3} + \dfrac{\sqrt{3}}{3} e \sin \sqrt{3} i + \dfrac{\sqrt{3}}{3} e \sin \sqrt{3} j + \dfrac{\sqrt{3}}{3} e \sin \sqrt{3} k$.

16. (a) It is clear that

$$i^j = e^{\ln i j} = e^{\frac{\pi}{2} i j} = e^{\frac{\pi}{2} k} = k.$$

Furthermore we have

(b) $-j$;   (c) i;

(d) $\cos \sqrt{\dfrac{\pi^2}{4} + \ln^2 \sqrt{2}} + \dfrac{\sin \sqrt{\frac{\pi^2}{4} + \ln^2 \sqrt{2}}}{\sqrt{\frac{\pi^2}{4} + \ln^2 \sqrt{2}}} \left(\dfrac{\pi \sqrt{2}}{4} i - \dfrac{\pi \sqrt{2}}{4} j + \ln \sqrt{2} k \right)$;

(e) $\cos \sqrt{\dfrac{\pi^2}{4} + \ln^2 \sqrt{2}} + \dfrac{\sin \sqrt{\frac{\pi^2}{4} + \ln^2 \sqrt{2}}}{\sqrt{\frac{\pi^2}{4} + \ln^2 \sqrt{2}}} \left(-\dfrac{\pi \sqrt{2}}{4} i + \ln \sqrt{2} j + \dfrac{\pi \sqrt{2}}{4} k \right)$;

(f) $\cos \sqrt{\dfrac{\pi^2}{4} + \ln^2 \sqrt{2}} + \dfrac{\sin \sqrt{\frac{\pi^2}{4} + \ln^2 \sqrt{2}}}{\sqrt{\frac{\pi^2}{4} + \ln^2 \sqrt{2}}} \left(\ln \sqrt{2} i + \dfrac{\pi \sqrt{2}}{4} j - \dfrac{\pi \sqrt{2}}{4} k \right)$.

17. Step by step we obtain

$$j^k = i, \quad j^k(1 + i - j) = -1 + i - k, \quad \overline{2 - i - 3j + k} = 2 + i + 3j - k,$$

$$[i + j, i - j]$$

$$= \mathrm{Sc}(i + j)\mathrm{Sc}(i - j) - \mathrm{Vec}(i + j)\mathrm{Vec}(i - j) + \mathrm{Sc}(i + j)\mathrm{Vec}(i - j)$$

$$+ \mathrm{Sc}(i - j)\mathrm{Vec}(i + j) = 0.$$

Then the given quaternion can be reduced to $1 + 2i + 3j - 2k$. Its first matrix representation is

$$\begin{pmatrix} 1 + 2i & -2 + 3i \\ 2 + 3i & 1 - 2i \end{pmatrix}.$$

18. Step by step we obtain

$$j^k = i, \quad \frac{1 - i - k}{j + k} = \frac{(1 - i - k)(-j - k)}{(j + k)(-j - k)} = \frac{-1 - i - 2j}{2},$$

$$j^k \frac{1 - i - k}{j + k} = i\left(-\frac{1}{2} - \frac{1}{2}i - j\right) = \frac{1}{2} - \frac{1}{2}i - k.$$

19. Let

$$A_1 = \ln \sqrt{a_1^2 + b_1^2 + c_1^2 + d_1^2}, \quad B_1 = \frac{b_1 \arccos \frac{a_1}{\sqrt{a_1^2 + b_1^2 + c_1^2 + d_1^2}}}{\sqrt{b_1^2 + c_1^2 + d_1^2}},$$

$$C_1 = \frac{c_1 \arccos \frac{a_1}{\sqrt{a_1^2 + b_1^2 + c_1^2 + d_1^2}}}{\sqrt{b_1^2 + c_1^2 + d_1^2}}, \quad D_1 = \frac{d_1 \arccos \frac{a_1}{\sqrt{a_1^2 + b_1^2 + c_1^2 + d_1^2}}}{\sqrt{b_1^2 + c_1^2 + d_1^2}},$$

$$A_2 = A_1 a_2 - B_1 b_2 - C_1 c_2 - D_1 d_2, \quad B_2 = A_1 b_2 + B_1 a_2 + C_1 d_2 - D_1 c_2,$$

$$C_2 = A_1 c_2 - B_1 d_2 + C_1 a_2 + D_1 b_2, \quad D_2 = A_1 d_2 + B_1 c_2 - C_1 b_2 + D_1 a_2.$$

Then we have

$$\ln(p) = A_1 + B_1 i + C_1 j + D_1 k,$$

$$\ln(p)q = (A_1 + B_1 i + C_1 j + D_1 k)(a_2 + b_2 i + c_2 j + d_2 k)$$
$$= A_2 + B_2 i + C_2 j + D_2 k,$$

and so

$$p^q = e^{\ln(p)q}$$

$$= e^{\sqrt{A_2^2 + B_2^2 + C_2^2 + D_2^2}} \times$$

$$\times \left(\cos \sqrt{B_2^2 + C_2^2 + D_2^2} + \frac{B_2 i + C_2 j + D_2 k}{\sqrt{B_2^2 + C_2^2 + D_2^2}} \sin \sqrt{B_2^2 + C_2^2 + D_2^2} \right).$$

20. We have

$$i^j = k, \quad i^k = -j, \quad j^k = i, \quad j^i = -k, \quad k^i = j, \quad k^j = -i,$$

so $A = 0$.

Chapter 6

1. (a) We have

$$\mathrm{Sc}(i) = 0, \quad \mathrm{Vec}(i) = (1,0,0) = i, \quad |\mathrm{Vec}(i)| = 1, \quad \sin(i) = \sinh(1)i.$$

The other numerical results are: (b) $\sinh(1)j$;   (c) $\sinh(1)k$;   (d) $\cosh(1)$;
(e) $\cosh(1)$;   (f) $\cosh(1)$;   (g) $\tanh(1)i$;   (h) $\tanh(1)j$;   (i) $\tanh(1)k$.

2. (a) Set $p = 1 + i - j + k$. Then we get

$$p_0 = 1, \quad \mathbf{p} = (1,-1,1) = i - j + k, \quad |\mathbf{p}| = \sqrt{3},$$

$$\mathrm{sgn}(\mathbf{p}) = \frac{\sqrt{3}}{3}i - \frac{\sqrt{3}}{3}j + \frac{\sqrt{3}}{3}k,$$

$$\sin(1 + i - j + k) = \sin(1)\cosh(\sqrt{3}) + \frac{\sqrt{3}}{3}\cos(1)\sinh(\sqrt{3})i$$

$$- \cos(1)\frac{\sqrt{3}}{3}\sinh(\sqrt{3})j + \frac{\sqrt{3}}{3}\cos(1)\sinh(\sqrt{3})k.$$

The other cases follow: (b) $\cos(1)\cosh(1) + \sin(1)\sinh(1)k$;   (c) $\sin(1)$
$\cosh(1) - \cos(1)\sinh(1)j$;   (d) $\sin(1)\cosh(\sqrt{2}) + \frac{\sqrt{2}}{2}\cos(1)\sinh(\sqrt{2})i -$
$\frac{\sqrt{2}}{2}\cos(1)\sinh(\sqrt{2})j$.

3. (a) We have

$$\sin(i) = \sinh(1)i, \quad \cos(i) = \cosh(1), \quad \tan(i) = \tanh(1)i,$$

$$\cos(i + j) = \cosh(\sqrt{2}), \quad \sin(k) = (\sinh 1)k,$$

$$\tan(i) + \cos(i + j) + \sin(k) = \cosh(\sqrt{2}) + \tanh(1)i + \sinh(1)k;$$

(b) $\cos(\sqrt{2}) + \frac{i}{2}\left(\sinh(2) + \sqrt{2}\sinh(\sqrt{2})\right) + \frac{\sqrt{2}}{2}\sinh(\sqrt{2})j$.

4. Let $p = a + bi + cj + dk$ be a solution to the given equation. (a) Using the
previous exercise we obtain the following system for a, b, c, d:

$$\sin a \cosh\sqrt{b^2 + c^2 + d^2} = 0, \quad \frac{b\cos a \sinh\sqrt{b^2 + c^2 + d^2}}{\sqrt{b^2 + c^2 + d^2}} = 1,$$

$$\frac{c\cos a \sinh\sqrt{b^2 + c^2 + d^2}}{\sqrt{b^2 + c^2 + d^2}} = 0, \quad \frac{d\cos a \sinh\sqrt{b^2 + c^2 + d^2}}{\sqrt{b^2 + c^2 + d^2}} = 0.$$

Therefore,

$$\sin a = 0, \quad b \cos a \sinh \sqrt{b^2 + c^2 + d^2} = \sqrt{b^2 + c^2 + d^2},$$

$$c \cos a \sinh \sqrt{b^2 + c^2 + d^2} = 0,$$

$$d \cos a \sinh \sqrt{b^2 + c^2 + d^2} = 0.$$

It follows that

$$a = l\pi, \quad l \in \mathbb{Z},$$

$$b(-1)^l \sinh \sqrt{b^2 + c^2 + d^2} = \sqrt{b^2 + c^2 + d^2},$$

$$c \sinh \sqrt{b^2 + c^2 + d^2} = 0, \quad d \sinh \sqrt{b^2 + c^2 + d^2} = 0.$$

Firstly, let $\sqrt{b^2 + c^2 + d^2} = 0$, then $b = c = d = 0$, which is a contradiction, since $\sin(l\pi) = 0 \neq i$. Consequently, $\sqrt{b^2 + c^2 + d^2} \neq 0$. If $\sqrt{b^2 + c^2 + d^2} \neq 0$, then $c = d = 0$. Finally we have the system

$$a = l\pi, \quad b(-1)^l \sinh |b| = |b|, \quad c = d = 0,$$

which yields

$$p = l\pi + \ln(\sqrt{2} + (-1)^l)i, \quad l \in \mathbb{Z}.$$

The further results, with $l \in \mathbb{Z}$, are: (b) $p = l\pi + \ln(\sqrt{2} + (-1)^l)j$; (c) $p = l\pi + \ln(\sqrt{2} + (-1)^l)k$; (d) $p = (2l + 1)\frac{\pi}{2} + \ln(\sqrt{2} + (-1)^l)i$; (e) $p = (2l+1)\frac{\pi}{2} + \ln(\sqrt{2} + (-1)^l)j$; (f) $p = (2l+1)\frac{\pi}{2} + \ln(\sqrt{2} + (-1)^l)k$; (g)–(i) no solutions.

6. This result is not true in general. For instance, the equality is not valid for $p = \sqrt{3} + i$.

7. Since p is a pure quaternion, we have that

$$\sin(p) = \operatorname{sgn}(p) \sinh(|p|), \quad \cos(p) = \cosh(|p|).$$

Using that

$$\operatorname{sgn}^2(p) = -|p|^2 = -1,$$

we obtain that

$$\sin^2 p + \cos^2 p = 1_{\mathbb{H}}, \quad \cos^2 p - \sin^2 p = \sinh(2|p|).$$

After we add (a) and (b) we obtain (c). From (a) we subtract (b), and we obtain (d).

9. (a) no solutions; (b) no solutions; (c) every pure quaternion p for which it holds $|p| = \frac{1+\sqrt{2}}{2}$; (d)–(j) No solutions.

10. (a), (b), (d)–(j) no solutions; (c) every pure quaternion p for which it holds $|p| = \frac{1+\sqrt{2}}{2}$.

11. (a), (b), (d)–(j) no solutions; (c) Every pure quaternion p for which $|p| = 1$.

Chapter 7

1. (a) Clearly we have

$$\mathrm{Sc}(i) = 0, \quad \mathrm{Vec}(i) = i, \quad |\mathrm{Vec}(i)| = 1, \quad \sinh(i) = \sin(1)i.$$

For the remaining suproblems we get

(b) $\sin(1)j$; (c) $\sin(1)k$; (d) $\cos(1)$; (e) $\cos(1)$;

(f) $\cos(1)$; (g) $\tanh(1)i$; (h) $\tanh(1)j$;

(i) $\tanh(1)k$; (j) $\sinh(1)\cos\sqrt{3} + \dfrac{\sqrt{3}}{3}\sin(\sqrt{3})\cosh(1)i -$

$$-\frac{\sqrt{3}}{3}\sin(\sqrt{3})\cosh(1)j - \frac{\sqrt{3}}{3}\sin(\sqrt{3})\cosh(1)k;$$

(k) $\cosh(1)\cos(\sqrt{3}) + \dfrac{\sqrt{3}}{3}\sinh(1)\sin(\sqrt{3})i -$

$$-\frac{\sqrt{3}}{3}\sinh(1)\sin(\sqrt{3})j - \frac{\sqrt{3}}{3}\sinh(1)\sin(\sqrt{3})k.$$

2. (a) $\dfrac{i+j+k}{\sqrt{3}}\sin\sqrt{3}$; (b) $\cos\sqrt{3}$.

3. (a) $\sinh(1)\cos(\sqrt{3}) + \dfrac{\cosh(1)\sin(\sqrt{3})}{\sqrt{3}}(i + j + k)$; (b) $\cosh(1)\cos(\sqrt{3}) +$

$+ \dfrac{\sinh(1)\sin(\sqrt{3})}{\sqrt{3}}(i + j + k)$.

4. (a) Let $p = a + bi + cj + dk$. Using the previous exercise we obtain for a, b, c, d the system

$$\sinh(a)\cos\sqrt{b^2 + c^2 + d^2} = 0, \qquad \frac{b\cosh(a)\sin\sqrt{b^2 + c^2 + d^2}}{\sqrt{b^2 + c^2 + d^2}} = 1,$$

$$\frac{c\cosh(a)\sin\sqrt{b^2 + c^2 + d^2}}{\sqrt{b^2 + c^2 + d^2}} = 0, \qquad \frac{d\cosh(a)\sin\sqrt{b^2 + c^2 + d^2}}{\sqrt{b^2 + c^2 + d^2}} = 0.$$

First set $b = c = d = 0$. Then $\sinh(a) = 0$ i.e., $a = 0$. This is a contradiction, since $\sinh(0) \neq i$. Now let $c = d = 0$. For a and b we get the system

$$\sinh(a)\cos|b| = 0, \quad b\cosh(a)\sin|b| = |b|.$$

Therefore,

$$a = 0, \quad b = (4l + 1)\frac{\pi}{2}, \quad l \in \mathbb{Z}.$$

Consequently, $p = (4l + 1)\frac{\pi}{2}i$, $l \in \mathbb{Z}$. The solutions for the equations (b)–(f) are given by

(b) $(4l + 1)\frac{\pi}{2}i, \quad j \in \mathbb{Z}$, (c) $(4l + 1)\frac{\pi}{2}i, \quad k \in \mathbb{Z}$,

(d) $\ln(\sqrt{2} + (-1)^l) + (2l + 1)\frac{\pi}{2}i, \quad \ln(\sqrt{2} + (-1)^{l+1}) - (2l + 1)\frac{\pi}{2}i$,

(e) $\ln(\sqrt{2} + (-1)^l) + (2l + 1)\frac{\pi}{2}j, \quad \ln(\sqrt{2} + (-1)^{l+1}) - (2l + 1)\frac{\pi}{2}j$,

(f) $\ln(\sqrt{2} + (-1)^l) + (2l + 1)\frac{\pi}{2}k, \quad \ln(\sqrt{2} + (-1)^{l+1}) - (2l + 1)\frac{\pi}{2}k$.

5. The solutions are given by

(a) $p = \ln\sqrt{2 + \sqrt{3}} + \frac{\sqrt{3}}{3}\arccos\sqrt{2}(i + j + k)$;

(b) $p = a = \frac{(2l+1)\pi\sqrt{6}}{6}(i - j + k), \quad l \in \mathbb{Z}, \quad \sinh(a) = (-1)^l\frac{\sqrt{6}}{2}$.

6. (a) We find step by step

$$pq = (1 + i - j - k)(1 + i) = 2i - 2j, \quad p_0 = 1,$$

$$\mathbf{p} = i - j - k = (1, -1, -1), \quad q_0 = 1, \quad \mathbf{q} = i = (1, 0, 0),$$

$$(p, q) = p_0\mathbf{q} - q_0\mathbf{p} - \mathbf{p} \times \mathbf{q} = i + j,$$

$$\sinh(k) = \sin(1)k, \quad \bar{r} = -k, \quad \bar{r}q = -k(1 + i) = -k - j,$$

$$q \cdot r = q_0 r_0 + \mathbf{q}\mathbf{r} = 0.$$

Therefore,

$$pq - (p, q) + \sinh(r) - \bar{r}q + q \cdot r = i - 2j + k(1 + \sin(1)).$$

For the second problem we have: (b) $1 + \frac{\pi}{3} + \ln\sqrt{2} + \frac{2\pi}{3}i + k\sin(1)$.

7. (a) $\cos\left(\frac{\pi}{2}\sin(1)\right) - j\sin\left(\frac{\pi}{2}\sin(1)\right)$;   (b) $\cos\left(\frac{\pi}{2}\sin(1)\right) + k\sin\left(\frac{\pi}{2}\sin(1)\right)$;
   (c) $\cos\left(\frac{\pi}{2}\sin(1)\right) - k\sin\left(\frac{\pi}{2}\sin(1)\right)$;   (d) $\cos\left(\frac{\pi}{2}\sin(1)\right) + i\sin\left(\frac{\pi}{2}\sin(1)\right)$;
   (e) $\cos\left(\frac{\pi}{2}\sin(1)\right) + j\sin\left(\frac{\pi}{2}\sin(1)\right)$;   (f) $\cos\left(\frac{\pi}{2}\sin(1)\right) - i\sin\left(\frac{\pi}{2}\sin(1)\right)$.

9. Since $\lim_{p\to 0_{\mathbb{H}}} \frac{\sin|p|}{|p|} = 1_{\mathbb{H}}$ and $\sinh(p) = \frac{p}{|p|}\sin|p|$, we conclude that

$$\lim_{p\to 0_{\mathbb{H}}} \sinh(p) = 0_{\mathbb{H}}.$$

10. We have that $\cosh(p) = \cos|p|$, so $\lim_{p\to 0_{\mathbb{H}}}\cosh(p) = 1_{\mathbb{H}}$.

11. Since $\sinh^2(p) = -\sin^2|p|$, $\cosh^2(p) = \cos^2|p|$, it follows that

$$\cosh^2 p - \sinh^2 p = \cos^2|p| + \sin^2|p| = 1_{\mathbb{H}}.$$

Chapter 8

1. (a) From the definition we can easily find the relations

$$\operatorname{arcsinh}(1+i) = \ln\left(1 + i + \sqrt{1+2i}\right), \quad \sqrt{1+2i} = e^{\frac{1}{2}\ln(1+2i)},$$

$$\ln(1+2i) = \ln\sqrt{5} + i\arccos\frac{\sqrt{5}}{5}, \quad \sqrt{1+2i} = \sqrt{\frac{\sqrt{5}+1}{2}} + i\sqrt{\frac{\sqrt{5}-1}{2}},$$

and therefore

$$\operatorname{arcsinh}(1+i) = \ln\sqrt{2 + \sqrt{5} + 2\sqrt{\frac{\sqrt{5}+1}{2}} + 2\sqrt{\frac{\sqrt{5}-1}{2}}}$$

$$+ i\arccos\frac{1 + \sqrt{\frac{\sqrt{5}+1}{2}}}{\sqrt{2 + \sqrt{5} + 2\sqrt{\frac{\sqrt{5}+1}{2}} + 2\sqrt{\frac{\sqrt{5}-1}{2}}}}.$$

The results for the problems (b)–(j) are: (b) $\frac{\pi}{2}i$;   (c) $\frac{\pi}{2}j$;   (d) $\frac{\pi}{2}k$;
(e) $\ln\left(1 + \sqrt{2}\right) + \frac{\pi}{2}i$;   (f) $\ln\left(1 + \sqrt{2}\right) + \frac{\pi}{2}j$;   (g) $\ln\left(1 + \sqrt{2}\right) + \frac{\pi}{2}k$;
(h) $\frac{\pi}{4}i$;   (i) $\frac{\pi}{4}j$;   (j) $\frac{\pi}{4}k$.

2. (a) $\dfrac{\ln\sqrt{5}}{2} + \dfrac{\arccos\frac{2}{\sqrt{5}}+\frac{\pi}{2}}{2}i$;   (b) $\dfrac{\ln\sqrt{5}}{2} + \dfrac{\arccos\frac{2}{\sqrt{5}}+\frac{\pi}{2}}{2}j$;   (c) $\dfrac{\ln\sqrt{5}}{2} + \dfrac{\arccos\frac{2}{\sqrt{5}}+\frac{\pi}{2}}{2}k$;

(d) $\dfrac{1}{2}\ln\sqrt{\dfrac{7}{3}} + \dfrac{\arccos\frac{2}{\sqrt{7}}+\frac{\pi}{2}}{2\sqrt{3}}(i+j+k)$;   (e) $\dfrac{1}{2}\ln\sqrt{\dfrac{7}{3}} + \dfrac{\arccos\frac{2}{\sqrt{7}}+\frac{\pi}{2}}{2\sqrt{3}}(-i-j+k)$;

(f) $\dfrac{1}{2}\ln\sqrt{\dfrac{7}{3}} + \dfrac{\arccos\frac{2}{\sqrt{7}}+\frac{\pi}{2}}{2\sqrt{3}}(-i-j-k)$.

3. (a) We know that $\ln(i) = i\frac{\pi}{2}$ and $\sinh(i) = i\sin(1)$. Further, we have

$$\ln(1-i-j) = \ln\sqrt{3} - \frac{\sqrt{2}}{2}\arccos\frac{\sqrt{3}}{3}i - \frac{\sqrt{2}}{2}\arccos\frac{\sqrt{3}}{3}j,$$

$$\operatorname{arctanh}(1-i-j-k) = \frac{1}{2}\ln\sqrt{\frac{7}{3}} - \frac{i+j+k}{2\sqrt{3}}\left(\arccos\frac{2}{\sqrt{7}} + \frac{\pi}{2}\right),$$

which leads to

$$\ln(i) + \ln(1-i-j) + \sinh(i) + \operatorname{arctanh}(1-i-j-k)$$

$$= \ln\sqrt{3} + \frac{1}{2}\ln\sqrt{\frac{7}{3}} + i\left(\frac{\pi}{2} - \frac{\sqrt{2}}{2}\arccos\frac{\sqrt{3}}{3} - \frac{1}{2\sqrt{3}}\arccos\frac{2}{\sqrt{7}} - \frac{\pi}{4\sqrt{3}}\right)$$

$$+ j\left(-\frac{\sqrt{2}}{2}\arccos\frac{\sqrt{3}}{3} - \frac{1}{2\sqrt{3}}\arccos\frac{2}{\sqrt{7}} - \frac{\pi}{4\sqrt{3}}\right)$$

$$+ k\left(-\frac{1}{2\sqrt{3}}\arccos\frac{2}{\sqrt{7}} - \frac{\pi}{4\sqrt{3}}\right).$$

For the other problems we have

$$\text{(b)}\ \ln\sqrt{5} + \left(\arccos\frac{2}{\sqrt{5}} + \frac{\pi}{2}\right)i + \left(\frac{\pi^2}{4} + \sin^2 1\right)k;$$

$$\text{(c)}\ \frac{\ln\sqrt{5}}{2} + i\frac{\pi}{2} + \frac{\arccos\frac{2}{\sqrt{5}}+\frac{\pi}{2}}{2}i - \frac{\pi}{2}k.$$

4. (a) We have

$$\arcsin(i) = i\operatorname{arcsinh}(ii) = i\arcsin(-1) = -\frac{\pi}{2}i.$$

The remaining problems add up to (b) $-\frac{\pi}{2}j$;   (c) $-\frac{\pi}{2}k$;   (d) $-\frac{\pi}{2}+\ln(1+\sqrt{2})i$; (e) $-\frac{\pi}{2} + \ln(1+\sqrt{2})j$;   (f) $-\frac{\pi}{2} + \ln(1+\sqrt{2})k$.

Chapter 9

1.

$$\text{(a)}\ \begin{pmatrix} j & -k \\ -1 + j - \frac{1}{2}k & i \end{pmatrix};\qquad \text{(b)}\ \begin{pmatrix} j & k \\ -1 + j + \frac{1}{2}k & -i \end{pmatrix};$$

$$\text{(c)}\ \begin{pmatrix} -1 - i - j + k & j & -i & -1 \\ j & 1 + i + j + k & -i & -1 \\ 0 & -1 - i - j + 2k & j & -i \end{pmatrix};$$

$$\text{(d)}\ \begin{pmatrix} -1 + i + j + k & -j & i & -1 \\ -j & 1 - i - j + k & i & -1 \\ 0 & -1 + i + j + 2k & -j & i \end{pmatrix}.$$

2. It is easy to obtain

$$\begin{pmatrix} 1 - i - 4j & 1 + i - 6k \\ -j + 3k & 4 - 6j \end{pmatrix}.$$

3. We obtain

$$\text{(a)}\ \begin{pmatrix} j - k & i - k & 1 - k \\ 2 + 2i + j - k & -1 - i & 2j \\ -1 + i - j + k & -2 + j + k & -2 + 2i - 2k \end{pmatrix};\qquad \text{(b)}\ \begin{pmatrix} -2 + 2i \\ i + j + 2k \end{pmatrix}.$$

4. (a) We have

$$W = \begin{pmatrix} 0 & 0 \\ 0 & 1 \end{pmatrix} + \begin{pmatrix} 1 & -3i \\ -i & 1 + i \end{pmatrix} j.$$

Therefore,

$$W_1 = \begin{pmatrix} 0 & 0 \\ 0 & 1 \end{pmatrix},\qquad W_2 = \begin{pmatrix} 1 & -3i \\ -i & 1 + i \end{pmatrix},$$

whence

$$W_c = \begin{pmatrix} 0 & 0 & 1 & -3i \\ 0 & 1 & -i & 1 + i \\ -1 & -3i & 0 & 0 \\ -i & -1 + i & 0 & 1 \end{pmatrix}.$$

A direct computation shows that

$$W_c^{-1} = \begin{pmatrix} -\frac{1}{6} & -\frac{i}{6} & -\frac{1}{3}+\frac{1}{6}i & -\frac{1}{6}+\frac{2}{3}i \\ -\frac{i}{18} & \frac{1}{18} & -\frac{1}{18}+\frac{2}{9}i & -\frac{2}{9}-\frac{1}{18}i \\ \frac{1}{3}+\frac{1}{6}i & \frac{1}{6}+\frac{1}{3}i & \frac{1}{6} & \frac{1}{6}i \\ \frac{1}{18}+\frac{2}{9}i & \frac{2}{9}-\frac{1}{18}i & -\frac{1}{18}i & \frac{1}{18} \end{pmatrix}.$$

Consequently,

$$T_1 = \begin{pmatrix} -\frac{1}{6} & -\frac{i}{6} \\ -\frac{1}{18}i & \frac{1}{18} \end{pmatrix}, \quad T_2 = \begin{pmatrix} -\frac{1}{3}+\frac{1}{6}i & -\frac{1}{6}+\frac{2}{3}i \\ -\frac{1}{18}+\frac{2}{9}i & -\frac{2}{9}-\frac{1}{18}i \end{pmatrix}$$

and

$$W^{-1} = T_1 + T_2 j$$

$$= \begin{pmatrix} -\frac{1}{6} & -\frac{1}{6}i \\ -\frac{1}{18}i & \frac{1}{18} \end{pmatrix} + \begin{pmatrix} -\frac{1}{3}j+\frac{1}{6}k & -\frac{1}{6}j+\frac{2}{3}k \\ -\frac{1}{18}j+\frac{2}{9}k & -\frac{2}{9}j-\frac{1}{18}k \end{pmatrix}$$

$$= \begin{pmatrix} -\frac{1}{6}-\frac{1}{3}j+\frac{1}{6}k & -\frac{1}{6}i-\frac{1}{6}j+\frac{2}{3}k \\ -\frac{1}{18}i-\frac{1}{18}j+\frac{2}{9}k & \frac{1}{18}-\frac{2}{9}j-\frac{1}{18}k \end{pmatrix};$$

(b) W^{-1} doesn't exist;

$$(c)\ W^{-1} = \begin{pmatrix} \frac{1}{2}i-\frac{1}{2}k & \frac{1}{2}-\frac{1}{2}j \\ -\frac{1}{2}i+\frac{1}{2}k & -\frac{1}{2}-\frac{1}{2}j \end{pmatrix}.$$

5. (a) $(1\ \ 0)^T$;   (b) no solutions.
6. Use the results in [54].
7. Let $A \sim B$. Then one can find a matrix S such that S^{-1} exists and $S^{-1}AS = B$. Furthermore, let λ_r be a (right) characteristic root of A. Then there exists a nonzero vector X such that $AX = X\lambda$. Let $y = S^{-1}X$. Then

$$S^{-1}ASY = S^{-1}ASS^{-1}X = S^{-1}AX = S^{-1}X\lambda = Y\lambda.$$

8. (a) Let $x = a + bi + cj + dk,\ a,b,c,d \in \mathbb{R}$. Then

$$(1+i)x = a - b + (a+b)i + (c-d)j + (c+d)k,$$

$$xj = -c - di + aj + bk.$$

Thus, we obtain the following system for a, b, c, d:

$$\{a - b = -c, \quad a + b = -d, \quad c - d = a, \quad c + d = b.$$

It follows that $a = b = c = d = 0$. Consequently, $x = 0$. The remaining results are: (b) $x = 0$;   (c) $x = a + (a + d)j + dk, a, d \in \mathbb{R}$.

9. Let $x \in \mathbb{H}\backslash\{0\}$ be a solution to the equation $ax = xb$. We multiply this equation by x^{-1} from the left and obtain $b = x^{-1}ax$. Since $x \in \mathbb{H}\backslash\{0\}$, we have $a \sim b$. Hence,

$$a_0 = b_0, \quad |a| = |b|.$$

Now let $a_0 = b_0$ and $|\mathbf{a}| = |\mathbf{b}|$. Then $|a| = |b|$, and so $a \sim b$. Therefore there exists $x \in \mathbb{H}\backslash\{0\}$, such that $ax = xb$.

10. (a) no solution;   (b) no solution;   (c) it has a solution;   (d) no solution;   (e) no solution;   (f) no solution.
11. The following relations prove the result:

$$ah_1y = h_1ay = h_1yb, \quad ah_1yh_2 = h_1ybh_2 = h_1yh_2b.$$

Chapter 10

1. (a) $3iqjq(i + j)q - 2iqj$;   (b) $2iqjq(i - j)qj + qjqi + 2iqj$.
2. (a) 2;   (b) 3;   (c) 4.
3.

$$(a) \ (iqj + 1 + i)^2 = (iqj + 1 + i)(iqj + 1 + i)$$
$$= iqjiqj + iqj + iqji + iqj + 1 + i + iiqj + i + i^2$$
$$= -iqkqj + 2iqj - iqk - qj + 2i;$$

$$(b) \ -(1 + i)q(1 - j)q^2k - (1 + i)q(1 - j)qiqjqj - (1 + i)q(1 - j)q$$
$$+iqkqk + iqjqjqj + iqk;$$

$$(c) \ iqjqkqj + iqjq^2k + iqjqk + kq^2j - kqkqk + kq.$$

4. (a) We have

$$f(q) = -iqkqkqjqi - iqkqkqi - iqkq^2k - jqjqjqi - jqjqi + jqiqk.$$

Then

$$f(j) = 1 - k;$$

(b) $1 + j$.

5. (a) False;   (b) True;   (c) True;   (d) False.
6. Let $x = \frac{1}{2}c\,b^{-1}$. Then

$$a\frac{cb^{-1}}{2} + \frac{cb^{-1}b}{2} = \frac{c}{2} + \frac{c}{2} = c.$$

7. (a) Let $x = a + bi + cj + dk$, where $a, b, c, d \in \mathbb{R}$. Then

$$xj = -c - di + aj + bk,$$

$$x + xj = a - c + (b - d)i + (a + c)j + (b + d)k.$$

This yields the following system for a, b, c, d:

$$\begin{cases} a - c = 0, \\ b - d = 0, \\ a + c = 0, \\ b + d = 1. \end{cases}$$

Consequently,

$$a = 0, \quad b = \frac{1}{2}, \quad c = 0, \quad d = \frac{1}{2}.$$

Therefore, $x = \frac{1}{2}i + \frac{1}{2}k$ is a solution to the considered equation. (b) $x = \frac{3}{7} + \frac{2}{7}i - \frac{1}{14}j + \frac{5}{14}k$;   (c) $x = \frac{1}{2}j - \frac{1}{2}k$;   (d) no solutions;   (e) no solutions.
8. (a) Let $x = a + bi + cj + dk$, where $a, b, c, d \in \mathbb{R}$. Then

$$ix = -b + ai - dj + ck,$$

$$ixj = d - ci - bj + ak,$$

$$xk = -d + ci - bj + ak,$$

$$ix + ixj + xk = -b + ai + (-d - 2b)j + (c + 2a)k.$$

For a, b, c, d we obtain the system

$$\begin{cases} -b = 1, \\ a = 1, \\ -d - 2b = -1, \\ c + 2a = 1. \end{cases}$$

Therefore,

$$a = 1, \quad b = -1, \quad c = -1, \quad d = 3,$$

and so

$$x = 1 - i - j + 3k.$$

(b) $x = a + bi + aj + bk, \ a + b = \frac{1}{2}$; (c) $x = 1 - i - j$.

9. (a) We multiply the second equation from the right by k and obtain

$$\begin{cases} ix + yj = 1 + i + j, \\ -x - yj = 1 - i + j + k. \end{cases}$$

This yields

$$(-1 + i)x = 2 + 2j + k,$$

$$2x = (-1 + i)^{-1}(1 - i + j + k),$$

$$x = -1 - i - \frac{1}{2}j - \frac{3}{2}k.$$

From the first equation of the considered system we obtain

$$yj = 1 + i + j - ix,$$

$$yj = 2i - \frac{1}{2}j + \frac{1}{2}k.$$

Now we multiply the last equation from the right by j and obtain

$$y = -\frac{1}{2} + \frac{1}{2}i - 2k.$$

Consequently

$$\begin{cases} x = -1 - i - \frac{1}{2}j - \frac{3}{2}k, \\ y = -\frac{1}{2} + \frac{1}{2}i - 2k. \end{cases}$$

Also, for (b) we get

$$\begin{cases} x = \frac{1}{3} - \frac{2}{3}i + \frac{2}{3}k \\ y = -\frac{2}{3} - \frac{1}{3}i - \frac{1}{3}j. \end{cases}$$

10. (a) We multiply the first equation from the right by i and obtain

$$\begin{cases} y = (1+j)x + 1 - i + j, \\ kx + iyj + yk = 1 + j + k, \end{cases}$$

whence

$$(i + 2k)x + (1 + j)xk = -2i - k.$$

Now we look for a solution to the last equation in the form

$$x = a + bi + cj + dk, \quad a,b,c,d \in \mathbb{R}.$$

Therefore,

$$-3d + (2a - c)i + (-2d + b)j + (3a - c)k = -2i - k.$$

For a, b, c, d we have the system

$$\begin{cases} -3d = 0, \\ 2a - c = -2, \\ -2d + b = 0, \\ 3a - c = -1. \end{cases}$$

This yields

$$a = 1, \quad b = 0, \quad c = 4, \quad d = 0,$$

and so

$$\begin{cases} x = 1 + 4j \\ y = -2 - i + 6j. \end{cases}$$

Also, for (b) we have

$$\begin{cases} x = 1 - j - k, \\ y = 1 - k. \end{cases}$$

11. We multiply the second equation from the right by j and obtain

$$z = 1 + i - j - k + (1 - i)xj.$$

Now we multiply the first equation from the right by j and obtain

$$y = (1 + i)x(j + k) + 1 + k.$$

From here and from the third equation we have

$$xi - ixi + xj + xk - ixk = -1 + 3i.$$

Let $x = a + bi + cj + dk, a, b, c, d \in \mathbb{R}$. Then

$$-b - d + a + (a + b + c)i + (d - b + 2a)j + (a + 2b - c - d)k$$
$$= -1 + 3i.$$

For a, b, c and d we obtain the system

$$\begin{cases} -b - d + a = -1, \\ a + b + c = 3, \\ d - b + 2a = 0, \\ a + 2b - c - d = 0. \end{cases}$$

Therefore

$$a = \frac{2}{7}, \quad b = \frac{13}{14}, \quad c = \frac{25}{14}, \quad d = \frac{5}{14},$$

and

$$\begin{cases} x = \frac{2}{7} + \frac{13}{14}i + \frac{25}{14}j + \frac{5}{14}k, \\ y = -\frac{25}{7} - \frac{5}{7}i - \frac{13}{7}j + \frac{4}{7}k, \\ z = -\frac{3}{7} + \frac{17}{7}i + \frac{3}{14}j - \frac{5}{14}k. \end{cases}$$

12. (a) $(p + 2q)(p - q)$;   (b) $(2p + q)(p + q)$;   (c) $q(p + q)(p - 2q)$;
    (d) $(p - 3q)(p - 2q)p$;   (e) $(p - q)(p + q)^2$;   (f) $(p - q)(p + q)(2p - q)$;
    (g) $(2p - 3q)(p + q)(p - q)(p + 2q)$;   (h) $(p + q)(2p - q)(p + 3q)$;
    (i) $p^2(qp^3q + p^2qp^2)q$;   (j) $p(q + pqp^2)p$.
13. We have

$$f(\alpha_1) = \alpha_1^2 + (1 + i + j - k)\alpha_1 + \mu = 0_{\mathbb{H}},$$
$$f(\alpha_2) = \alpha_2^2 + (1 + i + j - k)\alpha_2 + \mu = 0_{\mathbb{H}},$$

whence

$$\alpha_1^2 + \alpha_2^2 + (1 + i + j - k)(\alpha_1 + \alpha_2) + 2\mu = 0_{\mathbb{H}}. \tag{11.2}$$

Also,

$$f(\alpha_1 + \alpha_2) = (\alpha_1 + \alpha_2)^2 + (1 + i + j - k)(\alpha_1 + \alpha_2) + \mu = 1 - i - j - k,$$

which implies,

$$\alpha_1^2 + \alpha_2^2 + (1 + i + j - k)(\alpha_1 + \alpha_2) + \mu = -2i - 2k,$$

From here and (11.3) we get $\mu = 2i + 2k$.

14. Since $\alpha_1, \alpha_2 \in \mathbb{H}$ are roots of the polynomial f, we have

$$\alpha_1^2 + \lambda\alpha_1 + \mu = 0_{\mathbb{H}},$$
$$\alpha_2^2 + \lambda\alpha_2 + \mu = 0_{\mathbb{H}},$$

whence

$$\alpha_1^2 - \alpha_2^2 + \lambda(\alpha_1 - \alpha_2) = 0_{\mathbb{H}}. \tag{11.3}$$

From $\alpha_1\alpha_2 = \alpha_2\alpha_1$ it follows that

$$\alpha_1^2 - \alpha_2^2 = (\alpha_1 + \alpha_2)(\alpha_1 - \alpha_2).$$

Using (11.3) and above equality we obtain

$$(\alpha_1 + \alpha_2 + \lambda)(\alpha_1 - \alpha_2) = 0_{\mathbb{H}},$$

which implies (since $\alpha_1 \neq \alpha_2$) that

$$\lambda = -1 + i + j + k.$$

Therefore

$$f(p) = p^2 + (-1 + i + j + k)p + \mu.$$

From $f(\alpha_1 + \alpha_2) = j + k$ we get $\mu = j + k$.

15. $\lambda = \frac{1 - 6i - 4j - 5k}{26}$.

16. $q(p) = ipj + p^2(1 + i + j + k) - p$.

17. $f(p) = ipjpk - p^2 - ip - kp + 1 + i + j + k$.

19. $\alpha = -2i + 2k$.

20. $\alpha = i - k$.

22.

(a) $$\sum_{k=0}^{2n} k \binom{p}{k}^{-1} = \left[\prod_{l=1}^{n} \frac{1}{l} \binom{p}{l} \right]^{-1} \sum_{m=1}^{n} \prod_{l=1, l \neq m}^{n} \frac{1}{l} \binom{p}{l};$$

(b) $$\sum_{k=0}^{2n} (-1)^l \binom{p}{k}^{-1} \binom{p}{2n-k}^{-1} =$$

$$= \left[\prod_{k=0}^{2n} (-1)^l \binom{p}{k} \binom{p}{2n-k} \right]^{-1} \sum_{m=0}^{2n} \prod_{l=0, l \neq m}^{2n} (-1)^l \binom{p}{l} \binom{p}{2n-l};$$

(c) $$\sum_{k=0}^{n} k \binom{p}{k}^{-2} = \left[\prod_{l=1}^{n} \frac{1}{l} \binom{p}{l}^2 \right]^{-1} \sum_{m=1}^{n} \prod_{l=1, l \neq m}^{n} \frac{1}{l} \binom{p}{l}^2.$$

Bibliography

1. K. Abdel-Khalek, Quaternion analysis (1996, peprint). arXiv:hepth/9607152
2. G. Aeberli, Der Zusammenhang zwischen quaternären quadratischen formen und idealen in quaternionenringen. Comment. Math. Helv. **33**, 212–239 (1959)
3. V. Bargmann, Über den Zusammenhang zwischen Semivektoren und Spinoren und die Reduktion der Diracgleichungen für Semivektoren. Helv. Phys. Acta **7**, 57–82 (1934)
4. I.Y. Baritzhack, R.R. Harman, Optimal fusion of a given quaternion with vector measurements. J. Guid. Control Dyn. **25**(1), 188–190 (2002)
5. W.E. Baylis, Why i? Am. J. Phys. **60**, 788–797 (1992)
6. B. Beck, Sur les équations polynomiales dans les quaternions. Enseign. Math. **25**(3–4), 193–201 (1979)
7. J. Brenner, Matrices of quaternions. Pac. J. Math. **1**, 329–335 (1951)
8. E. Cartan, Sur les groupes linéaires quaternioniens. Vierteljahrsschrift der Naturforschenden Gesellschaft in Zurich **85**, 191–203 (1940)
9. G. Combebiac, *Calcul des Triquaternions* (GauthierVillars, Paris, 1902)
10. S. De Leo, P. Rotelli, Representations of $U(1, q)$ and constructive quaternion tensor products. Nuovo Cimento B **110**, 33–51 (1995)
11. P. Du Val, *Homographies, Quaternions, and Rotations*. Clarendon Press, Oxford (1964)
12. S. Eilenberg, I. Niven, The "fundamental theorem of algebra" for quaternions. Bull. Am. Math. Soc. **50**, 244–248 (1944)
13. A. Einstein, W. Mayer, Die Diracgleichungen für Semivektoren. Proc. R. Acad. Amst. **56**, 497–516 (1933)
14. I. Gelfand, V. Retakh, R.L. Wilson, Quaternionic quasideterminants and determinants, in *Lie Groups and Symmetric Spaces*, ed. by F.I. Karpelevich, S.G. Gindikin. American Mathematical Society Translations Series 2, vol. 210 (American Mathematical Society, Providence, 2003), pp. 111–123
15. K. Gürlebeck, W. Sprößig, *Quaternionic Analysis and Elliptic Boundary Value Problems* (Birkhäuser, Basel, 1990)
16. K. Gürlebeck, W. Sprößig, *Quaternionic and Clifford Calculus for Physicists and Engineers* (Wiley, New York, 1997)
17. K. Gürlebeck, K. Habetha, W. Sprößig, *Holomorphic Functions in the Plane and n-Dimensional Space* (Birkhäuser, Basel, 2008)
18. F. Gürsey, *Applications of Quaternions to Field Equations*. Ph.D. thesis, University of London, 1950
19. W.R. Hamilton, On quaternions and the rotation of a solid body. Proc. R. Ir. Acad. **4**, 38–56 (1850)
20. W.R. Hamilton, *Elements of Quaternions*, Vol. I et II (First edition 1866; second edition edited and expanded by C.J. Joly 1899–1901; reprinted by Chelsea Publishing, New York, 1969)
21. T. Haslwanter, Mathematics of the three dimensional eye rotations. Vis. Res. **35**, 1727–1739 (1995)
22. A.S. Hathaway, Quaternion space. Trans. Am. Math. Soc. **3**, 46–59 (1902)

J.P. Morais et al., *Real Quaternionic Calculus Handbook*,
DOI 10.1007/978-3-0348-0622-0, © Springer Basel 2014

23. D. Hestenes, A unified language for mathematics and physics, in *Clifford Algebras and Their Applications in Mathematical Physics*, ed. by J.S.R. Chisholmand, A.K.Common (Reidel, Dordrecht, 1986), pp. 1–23
24. D. Janovská, G. Opfer, The classification and the computation of the zeros of quaternionic, two-sided polynomials. Numer. Math. **115**, 81–100 (2010)
25. D. Janovská, G. Opfer, A note on the computation of all zeros of simple quaternionic polynomials. SIAM J. Numer. Anal. **48**(1), 244–256 (2010)
26. T. Jiang, Cramer rule for quaternionic linear equations in quaternionic quantum theory. Rep. Math. Phys. **57**(3), 463–468 (2006)
27. T. Jiang, M. Wei, On a solution of the quaternion matrix equation $X - AXB = C$ and its application. Acta Math. Sin. **21**, 483–490 (2005)
28. W.J. Johnston, A quaternion substitute for the theory of tensors. Proc. R. Ir. Acad. A **37**, 13–27 (1926)
29. C.J. Joly, Quaternion invariants of linear vector functions and quaternions determinants. Proc. R. Ir. Acad. **4**, 1–15 (1896)
30. V. Kravchenko, M. Shapiro, *Integral Representations for Spatial Models of Mathematical Physics*. Research Notes in Mathematics (Pitman Advanced Publishing Program, London, 1996)
31. T.Y. Lam, *A First Course in Noncommutative Rings*. Graduate Texts in Mathematics, vol. 131 (Springer, New York, 1991)
32. J. Lambek, If Hamilton had prevailed: quaternions in physics. Math. Intell. **17**, 7–15 (1995)
33. H.C. Lee, Eigenvalues and canonical forms of matrices with quaternion coefficients. Proc. R. Ir. Acad. Sect. A **52**, 253–260 (1949)
34. M. Markic, Transformantes nouveau véhicule mathématique. Synthèse des triquaternions de Combebiac et du système géométrique de Grassmann Calcul des quadriquaternions. Ann. Fac. Sci. Toulouse, **28**, 103–148 (1936)
35. M. Markic, Transformantes nouveau véhicule mathématique. Synthèse des triquaternions de Combebiac et du système géométrique de Grassmann Calcul des quadriquaternions. Ann. Fac. Sci. Toulouse **1**, 201–248 (1937)
36. I. Niven, Equations in quaternions. Am. Math. Mon. **48**, 654–661 (1941)
37. R.E. O'Connor, G. Pall, The quaternion congruence $\bar{t}at = b(mod\ g)$. Am. J. Math. **61**(2), 487–508 (1939)
38. S. O'Donnel, *William Rowan Hamilton, Portrait of a Prodigy* (Boole Press, Dublin, 1983)
39. H.L. Olson, Doubly divisible quaternions. Ann. Math. **31**(3), 371–374 (1930)
40. G.B. Price, *An Introduction to Multicomplex Spaces and Functions* (Marcel Dekker, New York, 1991)
41. V. Retakh, R.L. Wilson, *Advanced Course on Quasideterminants and Universal Localization*. Notes of the Course (Centre de Recerca Matematica (CRM), Bellaterra, 2007)
42. M. Riesz, *Clifford Numbers and Spinors*. Lecture Series No. 38 (Institute for Fluid Dynamics and Applied Mathematics/University of Maryland, Baltimore, 1958). Reprinted in *Clifford Numbers and Spinors*, ed. by M. Riesz, E.F. Bolinder, P. Lounesto (Kluwer, Dordrecht, 1993)
43. W. Rindler, I. Robinson, A plainmans guide to bivectors, biquaternions, and the algebra and geometry of Lorentz transformations, in *On Einsteins Path Essays in Honour of Engelbert Schucking*, ed. by A. Harvey (Springer, New York, 1999), pp. 407–433
44. E. Sarrau, *Notions sur la théorie des quaternions*, Paris, Gauthier-Villars, 1889, 46 p.
45. R. Serôdio, L.-S. Siu, Zeros of quaternion polynomials. Appl. Math. Lett. **14**, 237–239 (2001)
46. A. Sudbery, Quaternion analysis. Math. Proc. Camb. Philos. Soc. **85**, 199–225 (1979)
47. P.G. Tait, On the linear and vector function. Proc. R. Soc. Edinb. **23**, 424–426 (1899)
48. P.G. Tait, On the claim recently made for Gauss to the invention (not the discovery) of quaternions. Proc. R. Soc. Edinb. **23**, 17–23 (1899/1900)
49. O. Teichmüller, Operatoren im Wachsschen Raum. J. Math. **174**, 73–124 (1935)
50. J.P. Ward, *Quaternions and Cayley Numbers* (Kluwer, Dordrecht, 1997)
51. W.H. Watson, On a system of functional dynamics and optics. Philos. Trans. R. Soc. A **236**, 155–190 (1937)

52. D. Weingarten, Complex symmetries of electrodynamics. Ann. Phys. **76**(2), 510–548 (1973)
53. P. Weiss, On some applications of quaternions to restricted relativity and classical radiation theory. Proc. R. Ir. Acad. A **46**, 129–168 (1941)
54. P. Weiss, An extension of Cauchy's integral formula bymeans of a Maxwell's stress tensor. J. Lond. Math. Soc. **21**(3), 210–218 (1946)
55. H. Weyl, Quantenmechanik und Gruppentheorie. Weyl, H. Publication: Zeitschrift für Physik **46**(1–2), 1–46 (1927)
56. N.A. Wiegmann, Some theorems on matrices with real quaternion elements. Can. J. Math. **7**, 191–201 (1955)
57. D.R. Wilkins, William Rowan Hamilton: mathematical genius. Phys. World **18**, 33–36 (2005)
58. J.G. Winans, Quaternion physical quantities. Found. Phys. **7**, 341–349 (1977)
59. L.A. Wolf, Similarity of matrices in which the elements are real quaternions. Bull. Am. Math. Soc. **42**, 737–743 (1936)
60. R.M.W. Wood, Quaternionic eigenvalues. Bull. Lond. Math. Soc. **17**, 137–138 (1985)
61. F. Zhang, *Permanent Inequalities and Quaternion Matrices*. Ph.D. dissertation, University of California at Santa Barbara, 1993
62. F. Zhang, On numerical range of normal matrices of quaternions. J. Math. Phys. Sci. **29**(6), 235–251 (1995)
63. F. Zhang, Quaternions and matrices of quaternions. Linear Algebra Appl. **251**, 21–57 (1997)

Index

J.P. Morais et al., *Real Quaternionic Calculus Handbook*,
DOI 10.1007/978-3-0348-0622-0, © Springer Basel 2014